Climate Protection and Development

Climate Protection and Development

Editors

Frank Ackerman
Richard Kozul-Wright
Rob Vos

BLOOMSBURY ACADEMIC

Published in association with the United Nations

First published in 2012 by:

Bloomsbury Academic

An imprint of Bloomsbury Publishing Plc
50 Bedford Square, London WC1B 3DP, UK
and
175 Fifth Avenue, New York, NY 10010, USA

CIP records for this book are available from the British Library and the Library of Congress.

ISBN 978-1-78093-169-2 (paperback)
ISBN 978-1-78093-173-9 (hardback)
ISBN 978-1-78093-170-8 (ebook)

This book is produced using paper that is made from wood grown in managed, sustainable forests. It is natural, renewable and recyclable. The logging and manufacturing processes conform to the environmental regulations of the country of origin.

Printed and bound in Great Britain by the MPG Books Group, Bodmin, Cornwall.

www.bloomsburyacademic.com

Note

The designations employed and the presentation of the material in this publication do not imply the expression of any opinion whatsoever on the part of the Secretariat of the United Nations concerning the legal status of any country or territory or of its authorities, or concerning the delimitations of its frontiers.

The term "country" as used in the text of the present report also refers, as appropriate, to territories or areas.

The designations of country groups in the text and the tables are intended solely for statistical or analytical convenience and do not necessarily express a judgement about the stage reached by a particular country or area in the development process.

Mention of the names of firms and commercial products does not imply the endorsement of the United Nations.

Symbols of United Nations documents are composed of capital letters combined with figures.

Contents

Acknowledgements

This book rests on work undertaken for the United Nations' annual flagship report World Economic and Social Survey 2009: Promoting Development, Saving the Planet, prepared by the Department of Economic and Social Affairs of the United Nations Secretariat (UN/DESA). Rob Vos and Richard Kozul-Wright led the core team that prepared the report. The core team included Imran Ahmad, Piergiuseppe Fortunato, Nazrul Islam, Alex Julca, Oliver Paddison and Mariangela Parra-Lancourt. Crucial inputs were received from Alex Izurieta of the Development Policy and Analysis Division; Tariq Banuri, David O'Connor, Chantal Line Carpentier and Fred Soltau of the Division for Sustainable Development; and Manuel Montes and Frank Schroeder, who at the time of writing were with the Financing for Development Office of UN/DESA.

Frank Ackerman of the Stockholm Environmental Institute (SEI) took care of preparing the abridged and updated version of the study as reflected in the present volume. Marion Davis of the SEI provided skilful editorial suggestions and took care of updating many of the time-sensitive data in the original report.

The editors are further grateful to Jan McAlpine and Barbara Tavora-Jainchill of the secretariat of the United Nations Forum on Forests for their inputs to the original versions of chapters III and VI. Inputs and comments to the original report were also gratefully received from funds and organizations across the wider United Nations system, including the Global Environment Facility, the International Finance Corporation, the International Labour Office (Employment Strategy Department), the United Nations Development Programme (Bureau for Development Policy, New York), the United Nations Environment Programme (Division of Technology, Industry and Economics, Paris), the United Nations Environment Programme Risø Centre (Copenhagen) and the United Nations Framework Convention on Climate Change Secretariat (Bonn).

Specific inputs were also received from researchers at the Australian National University, Tufts University and the University of Oregon and from the South Centre, Geneva. The analysis benefited from a number of background papers prepared especially for the original report by a number of prominent experts on climate change and development. Those background papers are available at http://www.un.org/en/development/desa/policy/wess/wess_bg_papers.shtml

This project would not have been possible without the overall guidance of Jomo Kwame Sundaram, Assistant Secretary-General for Economic Development of the United Nations, and without his encouragement to turn the report into a more widely accessible publication, this book may not have become a reality.

Finally, thanks also due to Valerian Monteiro for skilful typesetting of the book and to Leah C. Kennedy for editorial assistance.

About the Editors

Frank Ackerman is the Director of the Climate Economics Group of the US Center of the Stockholm Environmental Institute based at Tufts University in Somerville, Massachusetts. He is also a founder and steering committee member of Economics for Equity and Environment (the E3 Network) and a member scholar of the Center for Progressive Reform in Washington, D.C. He is also a senior research fellow at the Global Development and Environment Institute of Tufts University.

Richard Kozul-Wright is Chief of the Unit on Economic Cooperation and Integration among Developing Countries of UNCTAD, Geneva. At the time of writing he was chief of the Development Strategy and Policy Analysis Unit of the Development Policy and Analysis Division at the Department of Economic and Social Affairs of the United Nations Secretariat.

Rob Vos is the Director of the Development Policy and Analysis Division (DPAD) of the Department of Economic and Social Affairs of the United Nations Secretariat (UN/DESA), and Affiliated Professor of Finance and Development at the Institute of Social Studies of Erasmus University, The Hague.

About other Contributors

This book builds on work undertaken for the United Nations, ***World Economic and Social Survey 2009: Promoting Development, Saving the Planet***, prepared by staff of the UN's Department of Economic and Social Affairs. Rob Vos and Richard Kozul-Wright directed the research and were lead authors of the report. This revised, updated and abridged version was prepared with the support of **Frank Ackerman** as main editor. Next to the lead authors **Richard Kozul-Wright** and **Rob Vos**, the following authors contributed further to the original versions of the chapters:

Imran Ahmad, currently guest lecturer and doctoral scholar at the Australian National University; at the time of writing Senior Economic Affairs Officer of the Development Policy and Analysis Division of UN/DESA *(Chapter I)*

Tariq Banuri, Director of the Division for Sustainable Development of UN/DESA *(Chapter II)*

Chantal Line Carpentier, Senior Economic Affairs Officer of the Division for Sustainable Development of UN/DESA *(Chapter V)*

Piergiuseppe Fortunato, currently Economic Affairs Officer of the Unit on Economic Cooperation and Integration among Developing Countries, UNCTAD; at the time of writing Associate Economic Affairs Officer of the Development Policy and Analysis Division of UN/DESA *(Chapter V)*

S. Nazrul Islam, Senior Economic Affairs Officer of the Development Policy and Analysis Division of UN/DESA *(Chapter I)*

Alex Izurieta, Senior Economic Affairs Officer of the Development Policy and Analysis Division of UN/DESA *(Chapter I)*

Alex Julca, Economic Affairs Officer of the Development Policy and Analysis Division of UN/DESA *(Chapter III)*

Manuel Montes, currently Chief of Development Strategy and Policy Analysis Unit in the Development Policy and Analysis Division at UN/DESA; at the time of writing Chief of Policy Analysis and Development Branch, Financing for Development Office *(Chapter VI)*

David O'Connor, Chief of Branch of the Division for Sustainable Development of UN/DESA *(Chapter V)*

Oliver Paddison, Economic Affairs Officer of the Development Policy and Analysis Division of UN/DESA *(Chapter III)*

Mariangela Parra-Lancourt, currently Economic Affairs Officer of the Financing for Development Office of UN/DESA; at the time of writing Economic Affairs Officer of the Development Policy and Analysis Division of UN/DESA *(Chapter IV)*

Frank Schroeder, Economic Affairs Officer of the Financing for Development Office of UN/DESA *(Chapter VI)*

Fred Soltau, Economic Affairs Officer of the Division for Sustainable Development of UN/DESA *(Chapter II)*

Preface

The present book emanates from the research undertaken for the United Nations' 2009 *World Economic and Social Survey* on climate change and development. The key message of the report and retained in this book deserves widespread attention:

> The climate and development crises will be solved together, or not at all. And the faster we begin, the less painful—and more possible—the solutions will be.

Writing about climate change often falls into one of two opposite traps, both of which this book avoids. This is not a story of gloom and doom, of inevitable climate catastrophe. On the contrary, it spells out, in more detail than usual, what can and should be done to avert the real risks of disaster. But it is also not a story of complacent congratulation for "win-win" initiatives, cautiously incremental steps, and "green" consumer choices. It summons us to an endeavour worthy of the resources and ingenuity of the twenty-first century, to bold initiatives with big costs—and much bigger benefits.

Although emphasizing developing country perspectives, this book repeatedly evokes the transformative accomplishments of United States policies in the mid-twentieth century. At a time of troubles not unlike the present, the New Deal, the Tennessee Valley Authority and the Marshall Plan were initiatives that rescued the economy of the nation, and the world. The time has come for a global "new deal": a practical, forward-looking solution aimed at jointly averting a climate catastrophe and erasing the scourge of poverty. This book explores the interconnected issues of climate and development, laying the groundwork for just such a new deal. An overview of the six chapters that follow may provide a useful introduction.

Chapter I, "Climate change and the development challenge", explains why the two problems are inextricably linked. Climate change, caused above all by carbon dioxide emissions from fossil fuel combustion, threatens our common future; we are close to the threshold of causing dangerous, effectively irreversible changes in the earth's climate. The climate is a global public good; we all have a common interest in reducing the emissions that cause climate change.

Yet, it is abundantly clear that the minority of the world's population that lives in high-income, developed countries owes its high incomes to a history of fossil fuel-based industrialization. Because a significant fraction of carbon dioxide emissions remain airborne for centuries, the residue of emissions from past growth is still with us, using up most of the limited absorptive capacity of the atmosphere. This threatens to send an intolerable message to the lower-income majority of the world: the reason that you cannot use the existing, low-cost technologies of fossil fuel-based industrialization is that other people did so first.

Climate change must be addressed in a world in which most people's top priority is economic growth and development. While the climate is a global public good, economic development is not. We are all in the same boat, but some of us have much nicer cabins than others.

The only resolution to this dilemma is the creation of a low-emissions, high-growth scenario for development, based on new technologies that produce adequate energy and rising incomes with little or no greenhouse gas emissions. This is a tall order; it will require pushing against and expanding the technological frontier, and making the new technologies available to developing countries on affordable terms. It calls for a "big push" in investment, initially emphasizing public sector investment in infrastructure and energy, to stimulate complementary private sector investment. Financial support from developed countries will be needed to launch this effort, but the resulting high growth rates in developing countries will lead to a sharply reduced need for external resources in the near future.

Chapter II, "Climate mitigation and the energy challenge", looks at the changes that the low-emissions, high-growth scenario will require in the energy sector. A switch to renewable energy, enhancement of terrestrial carbon sinks through afforestation, and investment in energy efficiency will all be important. Studies by McKinsey & Company, among others, have identified extensive, low-cost abatement opportunities—if we start now. The longer we wait, the more expensive it will become to reduce emissions, in part because more new capital will be invested in old, high-emissions technologies.

Energy supply is essential to economic growth, and triggers forward and backward linkages throughout the economy, contributing to development. Yet today, energy use is very unequally distributed, with the equivalent of 100 to 300 kWh per person per day in developed countries, compared with 50 kWh or less in most developing countries, and 20 kWh or less in most of sub-Saharan Africa and South Asia. The challenge is to end energy poverty, bringing everyone up to a threshold of 100 kWh per day, without expanding dirty coal-burning power plants and other fossil fuel facilities. The large volume of investment in energy systems projected for developing countries in the next few decades makes it urgent to develop and provide access to clean energy technologies, to prevent investment in another generation of emissions-intensive coal plants.

Policies to promote clean energy, including feed-in tariffs and renewable portfolio standards, have been applied with some initial success in developed countries. These approaches may be less applicable in lower-income countries where fewer individuals and firms have the resources for rooftop or backyard installation of renewables. Research, development, and diffusion of new technologies will be needed to ensure the transformation of fast-growing energy systems in the developing world.

Chapter III, "The adaptation challenge", begins with the recognition that climate damages cannot entirely be avoided. Some climate change has already occurred, and more is in the pipeline, even under the most rapid abatement scenarios. Developing countries will, in most cases, be hit first and worst by climate impacts, due to geography (tropical, coastal, and island locations are particularly at risk) as well as economic conditions.

Adaptation to climate change is closely related to development: low-income countries often lack the resources needed to prepare in advance, and to recover after disasters; developed countries have the necessary resources for resilience, whether or not they use those resources wisely in practice. Conversely, the vulnerability of low-income countries often leads to climate-related losses that make development more difficult. Economic analysis is difficult because the needed adaptation measures are widely varied, based on local conditions; in contrast, mitigation often involves technologies that are applicable around the world.

An appropriate response involves "climate-smart" development, including adaptation measures that simultaneously meet other needs and avoid increasing vulnerability to other environmental risks. Excessive reliance on market forces and competition will not build the needed resilience. External funding will be required, perhaps in the range of $50 billion to $100 billion per year; although much smaller than the cost of emissions reduction, this is much larger than the amounts provided for adaptation through a scattered assortment of bilateral and multilateral funding mechanisms.

Chapter IV, "A state of change: climate and development policy", considers the national policies that are necessary to support the new approach to development proposed in this book. All the well-known success stories of past economic development have involved a burst of sustained growth, nurtured by a strong developmental state. The "big push" approach to investment advocated here is no different. A fast enough pace of investment will allow catching up to or even leapfrogging over existing technologies, increasing productivity, lowering costs, and stimulating linkages to other sectors. Phasing out dirty technologies, and diversifying away from agriculture and other climate-sensitive, resource-based industries, are essential goals for this development strategy.

There are many warnings about the risk of public investment crowding out private sector investors; but skillfully targeted public spending on energy, infrastructure, health and education can crowd in additional resources from the private sector—as demonstrated, for example, in Brazil's ethanol industry. There is a need for policies that foster deployment of new energy technologies, along with energy-efficient vehicles, buildings, and appliances. Coal is abundant and cheap in many developing countries; investment is badly needed in cleaner coal—if, for example, carbon capture and sequestration technology becomes practical at commercial scale—and lower-emission alternatives, principally renewables.

A broad range of public policies are needed to support the low-emissions, high-growth development pathway. Fiscal and monetary policies should give priority to increasing public spending in areas needed for the "big push" in investment. Industrial policy should include subsidized credits, loan guarantees, tax breaks and other measures to support private firms in targeted sectors. Market forces are important, but cannot by themselves guide the transition to climate-friendly economic growth.

Chapter V, "Technology transfer for climate protection", focuses on the problem of making new, low-emissions technologies available in the developing countries where they are so urgently needed. A rapid pace of investment often leads to technological upgrading, as the newest options are installed. Yet international trade and investment agreements have, in recent years, tilted heavily toward protection of intellectual property rights, threatening to slow the diffusion of new technologies. The Agreement on Trade-related Aspects of Intellectual Property Rights (TRIPS) and the broader World Trade Organization (WTO) framework of trade agreements seem to tightly constrain access to patented and other proprietary technologies. There is, however, some flexibility remaining in the language of these agreements, and it is possible that climate-friendly technologies might qualify for exceptions to the intellectual property rules, similar to those granted for sharing essential medicines, and for plant genetic resources.

There is a broad range of institutional arrangements that could be used to facilitate technology transfer. Open-source information sharing, perhaps in coordination with an international funding mechanism, is one option. Foreign direct investment (FDI) can transfer technologies, although its effects in practice depend on host country policies. The example of China shows that strong protection of intellectual property rights is not necessary to attract FDI. The Clean Development Mechanism (CDM), established under the Kyoto Protocol, is another option, although it has been too small and bureaucratic to accomplish much so far; an expanded, streamlined CDM could prove important in the future. Trade policy could affect technology; the risk of carbon "leakage" (production moving to countries with weaker climate policies to gain a competitive advantage), however, is often exaggerated.

New initiatives, needed for international capacity building, could include: a multilateral technology fund; a human skills transfer programme (perhaps including "reverse outsourcing," i.e., training programmes providing developed country expertise to teach skills in developing countries); and a public technology pool. New attempts to amend WTO rules, or to adopt climate-related waivers to existing rules, may be needed as well.

Chapter VI, "Financing the development response to climate change", addresses the costs and the funding mechanisms required to support the low-emissions, high-growth strategy for development. Estimates of global mitigation costs range up to 2 per cent of world output per year; estimates of adaptation costs are perhaps an order of magnitude smaller, although quite uncertain. Large fractions of these investments must occur in developing countries. Market-based scenarios often assume that price-based policies such as a cap-and-trade mechanism and carbon offsets will generate the bulk of the needed funding. In contrast, the scenario envisioned in this book starts with much larger upfront investments by developed countries—and leads to sufficiently rapid growth that developing countries can finance most of their ongoing investment needs before the middle of the century.

The one existing, large-scale application of carbon pricing occurs in the European Union's Emissions Trading System (ETS). Cap-and-trade systems such as ETS are designed around the needs of developed countries, and allow them to avoid mitigation at home by buying offsets abroad. A global carbon tax, while provoking intense political resistance and presenting complex problems of international harmonization, could provide a predictable price incentive and a significant source of funds. There will, however, be competing priorities for those funds. Other policies could include incentives for private funding, and public sector funding. The combination of private funds and developing-country public funds, however, will not be sufficient.

At present, international funding for climate investments in developing countries is fragmented in structure and inadequate in size. What is needed is a globally funded public investment programme to change countries' development trajectories and allow growth while reducing carbon dependence. It requires ambition on the scale of the Tennessee Valley Authority (TVA), a key part of the New Deal policies in the United States in the 1930s. The TVA transformed the economy of one of the poorest parts of the nation, creating jobs, clean energy, and environmental improvements. Launching the low-emissions, high-growth scenario requires a "global TVA", a similarly ambitious programme of investment in energy supply and infrastructure.

Estimates of total investment needs range up to $1 trillion per year in developing countries. This will come from a mix of public and private funds from domestic and international sources. There is a need, however, for international funding for a front-loaded, public sector commitment to achieve the "big push" in investment. The institutional framework for this funding should include a development accord addressing the issues of equity, development, and climate change; transparency in governance of financing, to avoid the restrictive conditions that have hampered some existing aid programmes; and a new governance structure to oversee the process, giving voice to both developed and developing countries. As Al Gore has suggested, the Marshall Plan provides a valuable model for meeting shared global challenges—wisdom that should now be applied to climate and development.

This book lays out a challenging agenda. Its forceful presentation of the needs and perspectives of developing countries may be unfamiliar or uncomfortable to readers in high-income countries. Yet it is essential that we all understand one another on these issues; it is unfortunately clear that any large country, or group of mid-sized countries, can veto any global climate solution by refusing to participate. A solution will only work if it works for everyone.

Sensible solutions are threatened not only by the strange persistence of science denial, but also by the strength of aggrieved self-interest, born perhaps of insecurity and competition in a stagnating economy and unravelling welfare state.

This book describes a new approach to climate and development, an alternative to business as usual that addresses both sustainability and equity. Solving both problems at once is, as this book suggests, the only approach that could work.

Frank Ackerman, Richard Kozul-Wright and Rob Vos
Boston, Geneva and New York, September 2011

Chapter I
Climate change and the development challenge

Introduction

We are living in the best and worst of times. Over the long sweep of history, our world has never been more prosperous, inventive or interconnected than it is today. Yet economic insecurity has become ubiquitous, social divisions are greater than ever, and the health of the planet has never been so fragile. These are interrelated challenges that can be effectively addressed only through cooperation and collective actions, at both the national and international levels.

In recent years, collective actions have been hampered by technocratic complacency, which privileged private means over public ends. Deregulation and corporate leadership were deemed to be all that was needed to find the quickest and most efficient solutions to contemporary policy challenges, from health care provision and urban renewal to poverty alleviation and climate change. This mindset has been surrounded by the rhetoric of targets, partnerships and synergies, which drains policy discussion of much of its substance and tends to gloss over the conflicts and difficult trade-offs involved in big policy challenges.

Climate change is not just a major challenge for the coming decades—it is an existential threat. Recent estimates suggest that 300,000 people die each year as a result of global warming, and the lives of 300 million more are seriously threatened. We know more than ever before about why this is happening. The Intergovernmental Panel on Climate Change (IPCC), established in 1988 by the United Nations Environmental Programme and the World Meteorological Organization, has provided invaluable information and analyses concerning why and how our climate is changing, and with what consequences. The wider scientific community has backed up their efforts with a mountain of supportive evidence and modelling exercises. Theirs is a sobering picture of how emissions of man-made greenhouse gases have already frayed our environmental fabric and threaten to rip it apart. The race to keep global temperatures within safe bounds is now a race against time. Global emissions have risen steadily in the last two decades, but to avoid potentially catastrophic impacts, a growing body of research indicates, emissions must start dropping before 2020, and be down to 50–60 per cent below 1990 levels by 2050—with continuing declines required thereafter.[1]

So far, knowledge of the science has not translated into a focused policy response. Although industrialized countries know that it is two centuries of their carbon-fuelled growth that underlies the warming trend, they have failed to commit the resources and political will needed to establish an alternative development pathway. At the same time, for most of the rest of the world, catching up through continued economic development remains a top priority. Given industrialized countries' inaction, it is difficult to persuade developing countries to find alternative (and expensive) energy sources to meet their own development objectives.

Hopes for a new round of climate negotiations have not yet been fulfilled; fundamental questions remain unanswered: How much emissions reduction should take place, and where and by when? How much will it cost to meet the targets, and how will they be covered? How should a proper and enhanced global adaptation response be framed in light of the significant impacts of climate change?

This book does not try to provide the answers—those can only be found through open, inclusive and frank negotiations among all the nations involved. But even assuming that an agreement is reached, the work of translating it into an effective programme of transformative change will require ongoing adjustments, continuous consultation and response to persistent policy challenges. Accordingly, our analysis looks at the building blocks of a long-term solution—mitigation, adaptation, technology and finance—in order to consider what is being asked of developing countries, and what the international community must do to ensure that they can meet those expectations without jeopardizing their development goals.

We proceed essentially by working back from 2050, by which time there are likely to be another 2 billion or more people on this planet,[2] the vast majority of whom will be living in the cities of the developing world. If current trends continue, not only will most of them be poor and insecure, but they will also be much more vulnerable to climate-related threats.

A crucial step toward the solution consists of lowering the level of emissions released into the atmosphere. This step is absolutely necessary—but it is not, alone, sufficient to achieve a sustainable solution to the crisis. In advanced countries, significant emissions reduction has to be accompanied by strong employment levels and a search for energy security. In developing countries, pursuit of a low-emissions path must be compatible with catch-up growth, industrialization and urban expansion.

Since this book focuses, to a great extent, on the interrelated climate and development challenges faced by the developing world, it pays particular attention to mitigation challenges around energy use (chap. II). But inasmuch as creating resilience to climate threats is essential for many poor countries, we seek to dispel the erroneous notion that countries must choose between mitigation and adaptation (chap. III). Thus we spell out the shared opportunities and synergies to be derived from investment-led responses to both these challenges, from forging truly integrated strategies, and from reviving the role of an effective developmental state (chap. IV).

The adjustments that are being asked of developing countries are unprecedented and will carry heavy investment costs, particularly in the initial stages of the transition. Those costs present the major obstacle to the development of low-emissions, high-growth pathways. But if properly managed, they can provide developing countries with a foundation for mobilizing their own resources to meet the climate challenge. Still, to ensure both sufficient technological transfers (chap. V) and sufficient access to financial resources (chap. VI) will require a level of international support and solidarity rarely mustered outside a wartime setting.

Development in a warming world

The development challenge

The industrial revolution, beginning in the late eighteenth century, inaugurated two processes of far-reaching consequences. The first enabled a select few countries to embark on a modern economic growth path, breaking the constraints on development imposed by the natural environment and the localization of economic activity. New levers of wealth creation emerged around market specialization, innovation and scale economies, and in the context of industrialization, urbanization and the greater interconnection of communities. In the wake of this transformation, the income gap between the group of early starters and the rest of the world widened rapidly, all the more since the exploitation of resources and markets by colonizers suppressed economic opportunities in many countries and communities across the world for a century or more.

The second process transformed the relation between humans and nature: instead of merely adapting, humans now dominated the environment and placed ever-increasing demands on it in the service of expanding output. In particular, the traditional energy sources (biomass, water and wind) used to complement manual labour and animal transport were replaced, initially by coal and then by oil, for the purpose of powering increasingly sophisticated machines and means of transport. Access to these cheaper fossil fuels has been a critical stepping stone on all modern development pathways. However, the full cost of exploiting carbon-based fuels and other natural assets has often gone unrecorded.

Over the past 50 years, developing countries have tried to close the economic gaps created over the previous two centuries. The process has not been smooth, and external constraints and shocks have persistently upset efforts in many countries. While some developing countries, particularly those in East Asia, have been successful, they have been atypical (see figure I.1). In fact, beginning with the debt crisis of the late 1970s, economic constraints tightened and shocks intensified, which led to a fragmented and divergent pattern of global growth. The most notable success story has been China, whose uninterrupted growth over the past 30 years has been one of the drivers of the positive aggregate trends in the recent social and economic performance of the developing world. Between 2002 and 2008, unprecedented strong growth was registered almost everywhere, including in the least developed countries, reflecting, in part, the growing economic interactions among developing countries themselves. However, that phenomenon came

Figure I.1:
The income gap between G7 countries and selected regions, 1980-2007

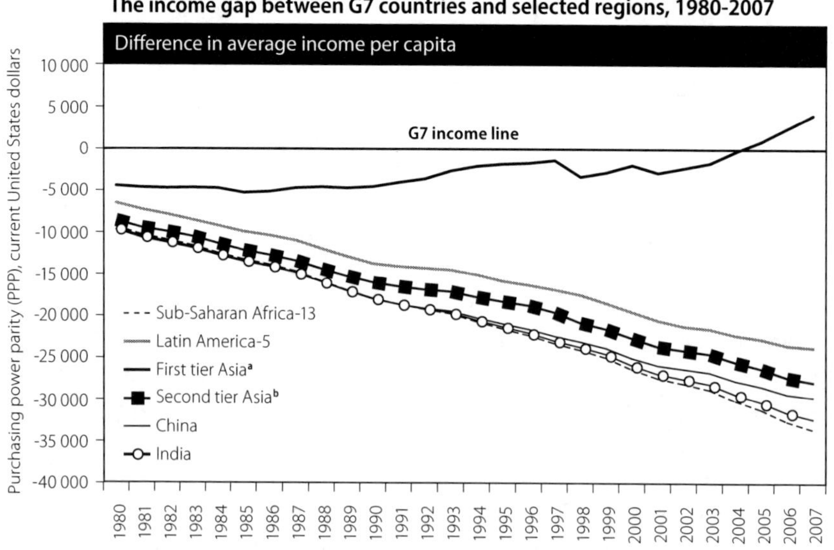

Source: UN/DESA calculations, based on WB-WDI online database.

a Hong Kong Special Administration Region of China; Republic of Korea, Singapore; Taiwan Province of China.

b Indonesia, Malaysia, Philippines, Thailand.

to an abrupt end with the onset of the most severe economic crisis since the 1930s. The heavy reliance on debt which fuelled much of that growth has proved an unreliable substitute for sound development strategy.

Government leaders in many developing countries are concerned that climate change is being used by those at the top of the development ladder to again hamper their nations' efforts to climb higher. How developing countries can achieve catch-up growth and economic convergence in a carbon-constrained world, and what the advanced countries must do to facilitate this, have become leading questions for policymakers at the national and international levels.

The climate challenge

The science is unequivocal: our climate is changing for the worse, owing to human activity. By causing an increase in emissions of greenhouse gases, human activity has led to an increase from the pre-industrial concentration of 280 parts per million (ppm) of carbon dioxide (CO_2) in the atmosphere, to more than 390 ppm today. This, in turn, is causing a major disruption in the natural climatic processes of the planet—and one that will not be eliminated overnight. Carbon dioxide and some other greenhouse gases have a long residence time in the atmosphere; in other words, once emitted, they remain there for decades.

Atmospheric carbon levels have reached such high levels primarily due to emissions from the production of energy for use by rich countries. Today fossil fuels are used to meet some 80 per cent of total energy needs. While energy emissions are the largest culprit by far, they are not the sole source of the problem (table I.1). Globally, forest ecosystems contained 652 billion tons of carbon in 2010, continuing a slow but steady decline over the last two decades;[3] changes in that mass of embodied carbon also have a large effect on atmospheric concentrations. Deforestation and forest degradation are the primary sources of carbon emissions from some developing countries. In 2005, forestry and land-use change released about 5.4 gigatons of CO_2 into the atmosphere, accounting for 15.9 per cent of all human-generated CO_2 emissions, and 12.2 per cent of total greenhouse gas emissions.[4]

The consequences of rising emission levels are now becoming clear. Global average surface temperature increased by almost 1° C between 1850 and 2000, with a noticeable acceleration in recent decades (see figure I.2). The global average sea level increased at an average rate of 1.8 millimetres (mm) per year over the last 50 years. More recently, from 1993 to 2008, however,

Table I.1
Greenhouse gas emissions (carbon dioxide, methane, nitrogen dioxide, perfluorocarbons, hydrochlorofluorocarbons and sulphur hexafluoride) by sector, 2005

Sector	*Megatons of CO_2 equivalent*	*Share (percentage)*
Energy	28,432	64.4
Electricity and heat	12,365	28.0
Manufacturing and construction	5,207	11.8
Transportation	5,374	12.2
Other fuel combustion	3,744	8.5
Fugitive emissions	1,743	4.0
Industrial processes	1,884	4.3
Agriculture	6,075	13.8
Land-use change and forestry	5,376	12.2
Waste	1,419	3.2
International bunkers	941	2.1
Total	**44,127**	**100.0**

Source: World Resources Institute, 2010.

Note: Details may not add to totals due to rounding.

Figure I.2:
Rise in global temperature, 1880-2010

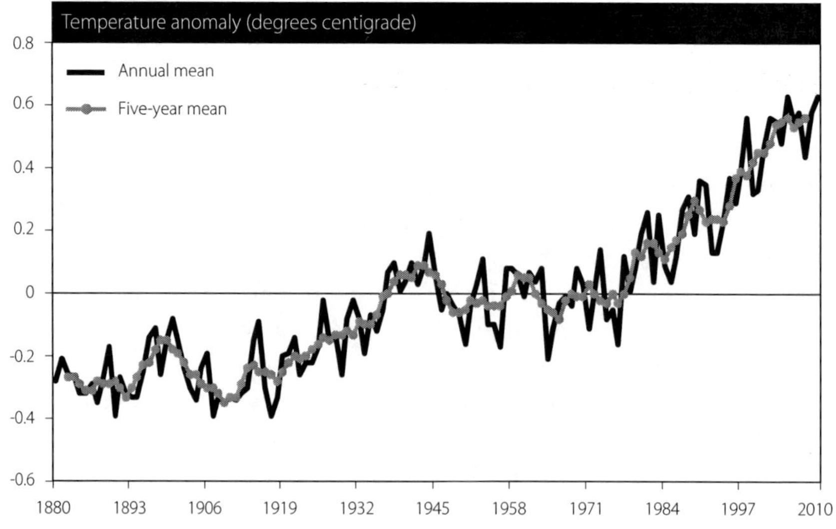

Source: National Aeronautics and Space Administration, Goddard Institute for Space Studies (GISS). Data available from http://www.data.giss.nasa.gov/gistemp/graphs.

it rose by 3.1 mm per year.[5] There have been large changes in patterns of precipitation, with significant increases in eastern parts of North and South America, Northern Europe, and northern and Central Asia, and decreases in the Sahel, the Mediterranean, Southern Africa and parts of South Asia. The area affected by drought has increased. Extreme weather events have increased in number, scope and intensity. Climate change is significantly affecting forests: there have been changes in their physiology, structure, species composition and health, largely owing to changes in temperature and rainfall. Many tropical forests in Latin America have experienced losses in biodiversity. Increased temperatures and drought result in more frequent outbreaks of pest infestations, more forest fires, and increasing alterations in populations of plant and animal species, severely affecting forest health and productivity.

A synthesis of the findings of a major climate science conference in March 2009, which included contributors to Intergovernmental Panel on Climate Change (IPCC) reports, notes:

> Many key climate indicators are already moving beyond the patterns of natural variability within which contemporary society and economy have developed and thrived. These indicators include global mean surface temperature, sea-level rise, global ocean temperature, Arctic sea ice extent, ocean acidification, and extreme climatic events. With unabated emissions, many trends in climate will likely accelerate, leading to an increasing risk of abrupt or irreversible climatic shifts.[6]

That the situation will worsen is no longer in doubt; the only question is by how much. Table I.2 presents the emission scenarios identified by the IPCC and their likely impact on temperatures and sea level by the end of this century. In general, greater emissions lead to greater changes in climate, while lower emissions lead to smaller changes. Moreover, as the IPCC has noted, its own scenarios report, released in 2000, and most scenarios developed since then fail to take into account the uncertainties regarding key climate processes and feedbacks. These include transmission of heat to lower depths of ocean, causing thermal expansion; contraction of the Greenland ice sheet; contraction of the western Antarctic ice sheet; reduction in the terrestrial and ocean uptake of atmospheric CO_2 as the CO_2 level rises; cloud feedback; and slowing down or even reversal of the meridional overturning circulation, among others. These feedbacks add another layer of complexity and uncertainty to future projections, but the uncertain risks largely point in the same direction, toward more severe or even catastrophic outcomes.

Table I.2
Emission scenarios and their impact

Scenario	Greenhouse gas concentration in 2100 (ppm CO_2e)	Temperature change (° C) in 2090-2099 relative to 1980-1999		Sea-level rise (metres) in 2090-2099 relative to 1980-1999
		Best estimate	*Likely range*	*Model-based range (excluding some future changes in ice flow)*
Constant year 2000 concentration		0.6	0.3-0.9	not available
B1	600	1.8	1.1-2.9	0.18-0.38
A1T	700	2.4	1.4-3.8	0.20-0.45
B2	800	2.4	1.4-3.8	0.20-0.43
A1B	850	2.8	1.7-4.4	0.21-0.48
A2	1250	3.4	2.0-5.4	0.23-0.51
A1FI	1550	4.0	2.4-6.4	0.26 0.59

Source: Intergovernmental Panel on Climate Change, 2007a, Table 3.1.

Note: The Intergovernmental Panel on Climate Change broadly identified four possible economic pathways (or "storylines"), referred to as A1 (a convergent world with fast economic growth); A2 (a non-convergent world with slow economic growth); B1 (a convergent and more environment-friendly world); and B2 (a non-convergent but environment-friendly world with an intermediate rate of economic growth). In addition to the above four broad storylines, the following three sub-variants of A1 have been distinguished, depending on the energy composition of economic growth: A1FI (relatively greater dependence on fossil fuels); A1B (a more balanced dependence on different energy sources); A1T (a greater reliance on non-fossil energy sources).

Many models also suggest that if emissions continue at "business as usual" rates, the stock of greenhouse gas emissions in the atmosphere will reach double the pre-industrial level in the second half of this century, resulting in a high probability of dangerous temperature rises, with potentially destabilizing economic and political consequences.

The interdependency challenge

The climate and development challenges are inextricably linked. When the overriding policy priority is economic growth, expanding the reach of energy and transportation infrastructure and making them available to an increasingly urban population and industrial workforce are unavoidable. So are major land-use changes. If developing countries simply stay on the path followed by today's rich countries, the impact on the earth's climate will be devastating.

At the same time, the prospects for more sustainable development are likely to be undermined by the direct and indirect impacts of climate change on economic growth. If climate damages reduce the resources available to invest in diversification and resilience, there will be heightened vulnerability to future climate trends and shocks. This vicious circle is already apparent in many arid and semi-arid countries in Africa. Adverse impacts on food and water supplies as well as on health conditions are likely to tighten growth constraints in other parts of the world.

An understanding of the complex ways in which economic development and climate variables interact is still evolving. However, the cumulative and unstable nature of that interaction poses obvious challenges for policymakers. This book seeks to build its assessment of that challenge around the pivotal role of investment and to examine some of the linkages and feedbacks that, from this starting point, can help define development strategies in a warming world.

From free-riding to burden-sharing

The Stern Review on the Economics of Climate Change, released by the British Government in October 2006, identified climate change as "the greatest market failure the world has seen" and provided a path-breaking attempt to model the cost of doing nothing compared with the cost of adopting an alternative strategy to hold emissions below a manageable threshold.[7] This perspective rests on a form of "climate ethics", centred on the challenge of providing a "global public good" and the need to realign social and private cost by making polluters pay for past, current and future damages from carbon emissions. The Stern Review concluded that future generations could be made much better off at relatively little cost to present generations.

Recognizing that a stable climate is a global public good makes an important point about the systemic nature of the challenge and the need for collective action to overcome it. Yet the interconnected problems of climate and development cannot be understood solely in these terms: while climate stability can only be provided to everyone at once, the same is clearly not true of economic development. There are intertwined problems of externalities, of vested interests and market power, and of uncertainty, making the market by itself an imperfect instrument for managing these challenges. Moreover, difficult distributional issues, rooted in a very uneven historical pattern of economic development, are obscured by the terminology of global public goods.

Historically speaking, it is largely emissions produced by the industrialized countries that have caused a rise in greenhouse gas concentrations. As table I.3 shows, developed countries produced three-quarters of the world's CO_2 emissions from 1850 to 2005, and the per capita emissions of most developing countries are much lower than those of developed nations. (The exceptions, such as Brazil in table I.3, are often countries with large emissions from deforestation and land use changes.)

Table I.3
Share of cumulative CO2 emissions, 1850-2005, and per capita emissions, 2005, selected countries

	Share of global cumulative CO_2 emissions 1850-2005 (percentage)	*Per capita emissions, 2005 (metric tons of CO_2e)*
Developed countries		
United States	29.2	23.5
European Union (27)	26.9	10.9
Russian Federation	8.2	14.2
Germany	7.1	12.2
United Kingdom	6.0	11.4
Japan	3.8	10.9
France	2.8	9.4
Canada	2.2	25.0
All Annex I	***74.6***	
Developing countries		
China	8.3	5.5
India	2.3	1.7
South Africa	1.1	9.2
Mexico	1.0	6.7
Brazil	0.8	15.3
Pakistan	0.2	1.6
Philippines	0.2	2.5
All non-Annex I	***25.4***	

Source: World Resources Institute, 2010. Annex I and non-Annex I figures calculated by the authors from CAIT 8.0 data.

Notes: Annex I grouping includes economies in transition. Global emissions 1850-2005 are for energy only. Per capita emissions include CO2, CH4, N2O, PFCs, HFCs, and SF6, including emissions from land-use change and forestry (LULUCF) and international bunkers.

Brazil's per capita emissions are a particularly stark example of the impact of the LULUCF category on some developing countries' emissions; without it, Brazil's per capita emissions are 5.5 tons for 2005.

Responsibility for emissions is a major factor in discussions of who should bear the costs of mitigation, so the question of whether to look at historical or current emissions is crucial. Much has been made recently, for example, of the fact that China has replaced the United States as the single largest emitter. However, China's *per capita* emissions levels remain far below those of the developed countries (and in fact, below those of many other developing countries); indeed, China's 2005 per capita emissions, 5.5 tons CO_2e, were still less than half those of the United States at the start of the First World War. On a cumulative historical basis, China is even farther from catching up to developed country levels (see figure I.3).

Wealthier countries' climate policies could also have adverse spillover effects for developing countries in terms of international trade, financial flows and commodity processes—as is already the case with the market impacts of the growth in biofuel use, for example. Industrialized countries' stance on intellectual property rights, meanwhile, will have a significant impact on technology transfer, which is crucial for developing countries (see chap. V). Against this backdrop, it seems unreasonable to accuse developing countries of wanting to be "free riders" in climate mitigation because they resist commitments being imposed on them. In fact, a much more nuanced framework will be needed to manage the burden of protecting the climate on an equitable basis. Several proposals for advancing the discussions are currently on the table (see box I.1).

Figure I.3:
Annual per capita emissions, selected regions, 1950-2008

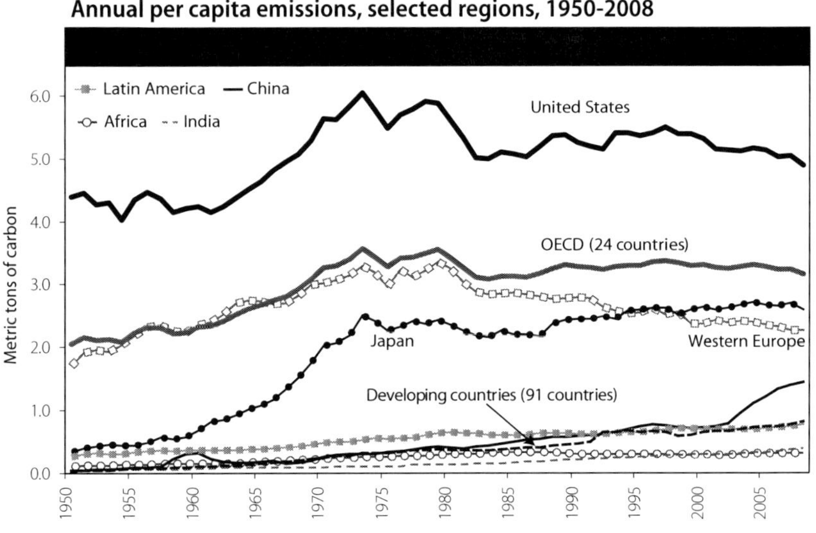

Source: See table I.3.

Box I.1: Burden-sharing proposals

Numerous burden-sharing mechanisms have been proposed by researchers and by participants in the global climate negotiation process, including:

Equal per capita emissions rights. A global limit on annual emissions is divided by world population, resulting in an equal per capita right to emit. Each country is allocated a level of emissions proportional to its population. The limit on global emissions would be reduced over time to achieve a desired stabilization trajectory (Agarwal and Narain 1991; Narain and Riddle 2007).

Individual targets. Again, a limit is set on world annual emissions, and a "universal cap" on per capita emissions is calculated. But then, to prevent high-emitting individuals from free-riding on low emitters' unused rights, each country's allocation is calculated as the sum of actual emissions for all residents with emissions below the cap, and the universal cap for all residents with emissions currently above the cap (Chakravarty et al. 2008).

Contraction and convergence. This plan phases in equal rights to emit by a target date. Each country is allocated emission rights based on its past emissions. Countries that exceed desired per capita global emissions have their allocation reduced in each succeeding year, while countries that emit less than this target receive a higher allocation each year. Over time, global emissions contract while high- and low- emitting countries converge on the same target per capita emissions (Global Commons Institute 2008).

One standard, two convergences. Each country gets an emission allowance based on the per capita emission reductions needed to meet a desired target. Different per capita ceilings are set for industrialized and developing countries, so developing countries have more room to grow their economies before having to decrease emissions to stay within their cumulative cap. The ceilings are adjusted annually until they achieve convergence. Trading of emissions rights makes it possible for all developing countries to use or benefit from their entire allowance (Gao 2007).

Greenhouse Development Rights. The burden of emissions reductions is allocated based on countries' relative capacity to pay and their responsibility for past and current emissions. Both criteria exempt per capita incomes and emissions below a "development threshold" in order to safeguard the right of low-income countries to economic growth, and to ensure that only individuals above the threshold income level have to pay for emissions abatement. Each country is assigned an obligation to pay for abatement—whether at home or abroad—based on its share of cumulative emissions since a base year (such as 1990) and the cumulative income of its population, counting only emissions and incomes above the development threshold (Baer, Athanasiou and Kartha 2007).

Revised Greenhouse Development Rights. Formulated by researchers at Tsinghua University in a report prepared by the Chinese Economists 50 Forum, this framework adjusts the GDRs model by including cumulative emissions going back to 1850, and accounting for emissions based on consumption (rather than on production) within each country. The result is a greater responsibility on the part of industrialized countries to pay for emissions reductions around the world (Fan et al. 2008).

Source: Ackerman and Stanton, 2009.

Still, in a very real sense, the future of the planet rests with the efforts of the developing world. Already, rich countries represent less than one sixth of the global population, and almost all of the additional billions of people to be added to that population over the next few decades will be in the developing world. Though industrialized countries will likely bear most of the initial costs of climate change mitigation, commensurate to their historical responsibility and greater economic resources, developing countries themselves will have to take measurable and verifiable steps to protect their own future.

Developing countries' scepticism about international mitigation efforts has been driven as much by developed countries' recent performance regarding climate policy as by their past record on economic development. For example, the Clean Development Mechanism, established under the Kyoto Protocol to the United Nations Framework Convention on Climate Change, was supposed to be an important link between developed countries' emission reduction efforts and efforts of developing countries, but has failed to live up to expectations, in terms of both quantity and quality. Similarly, the level of support provided to various funds set up to help developing countries adapt to climate change has been very low so far (see chaps. III and VI). Lack of bold and generous leadership has given rise to a lack of trust, which now represents a serious obstacle to mustering the international cooperation needed to deal effectively with the climate challenge.

Charles Kindleberger observed that in a world of interdependent nation states with widely differing access to economic resources and political power, effective multilateral cooperation depends on "positive leadership, backed by resources and readiness to make some sacrifice in the international interest".[8] He also recognized that the leadership role often goes unapplauded, particularly at home, and has a tendency to retreat or atrophy, but he argued that, particularly in a time of crisis, the hallmark of leadership is the willingness to assume responsibility. The urgency of the climate crisis certainly calls for a renewed leadership role from those countries most responsible.

International cooperation does not, however, hinge on leadership alone. Strong state capacities are needed, at all levels of development, to help shape a common and inclusive vision, to ensure that the limitations on national sovereignty in some areas are balanced by the opportunities opened up in others, and to guarantee effective participation in the negotiation and implementation of international rules, regulations and support mechanisms. In this respect, the erosion of state capacity in recent years, particularly in developing countries, represents an obstacle to international cooperation and has contributed to the lack of transparency and democratic accountability in many multilateral institutions. Correcting this is an urgent priority if real progress is to occur on the climate issue (for further discussion, see chap. IV).

The policy response

The timid policy response to the climate crisis to date reflects the fact that climate change has so far been a slow-moving process; its impacts have been less perceptible than many other shocks and crises confronting policymakers in the "normal" political cycle. Moreover, its consequences may have seemed easy for some to ignore, since their brunt has been borne by the poorest countries and communities.

Yet, climate scientists have continued to build a vast array of evidence and analyses demonstrating the unprecedented historical scale and speed of greenhouse gas increases, the signs of acceleration, the damage that has already been done to the climate, and the risks of getting locked into irreversible pathways if trends continue. This has led some countries to adopt ambitious targets for emissions reduction—but opinion surveys suggest that there is still a long way to go to convince politicians and the public of the urgency of the challenge.

The environmental movement has a long track record of not only warning of the dangers of unchecked pollution and reckless exploitation of natural assets, but also organizing successful campaigns on local issues. The political parties, regulatory reforms and environmental ministries that have emerged from those campaigns have taken the lead in responding at the national level to environmental challenges, especially in developed countries. Moreover, this community has been on the front lines in the ideological battle against climate sceptics. On the other hand, it has struggled to forge its own integrated perspective on the economic, political and scientific dimensions of the climate challenge, particularly in the international arena. Even where such a perspective has begun to emerge in richer countries, the implications for the developing world, where rapid growth, industrialization and urbanization remain paramount goals, have not been clearly or convincingly spelled out.

Economists have contributed their own perspectives on climate change, focusing on policy options. Theirs is a language of risk assessment, measured trade-offs between costs and benefits, marginal price changes, and discounting of future outcomes. Their "integrated assessment models" have an aura of quantitative rigour and precision, but they typically endorse an overly cautious approach, either by showing that a slow response is economically optimal, or by offering facile solutions to "externalities" so the market can reassume its central role.[9] In the context of climate change, their practical policy advice has focused on the mechanics of carbon taxes or trading schemes, and on the dangers of ambitious climate initiatives constraining future growth.

Integrated assessment models also generally have little to say about structural inequality or historical development; this has been a long-standing target of criticism of conventional economic models. Perhaps more surprising, however, has been the cavalier attitude of many economists towards climate risk. The Oxford economist Dieter Helm has argued that current climate policy and targets are being designed on the basis of current economic structures and of how marginal emissions reductions can be achieved from such a starting point, but with very little attention paid to long-term structural trends.[10] This approach is likely to seriously underestimate the size and cost of the challenge. One set of estimates of probabilities that scientists attach to the occurrence of higher temperatures are given in table I.4; they are far higher than risks that often lead individuals to take out insurance against worst-case scenarios. On this basis, some economists have suggested that the risks of catastrophic global warming merit a collective response that could be viewed as a planetary insurance policy.

Meanwhile, those who are shaping climate policy may not realize the implications of the adjustments they are asking of developing countries.

Table I.4
Probability of exceeding specified temperature increases at different greenhouse gas stabilization levels

Stabilization level (ppm CO_2e)	*Increase in temperature (°C above the pre-industrial level)*					
	2	**3**	**4**	**5**	**6**	**7**
450	78	18	3	1	0	0
500	96	44	11	3	1	0
550	99	69	24	7	2	1
650	100	94	58	24	9	4
750	100	99	82	47	22	9

Source: Stern, 2009, p. 26, citing Murphy et al., 2004.
Note: The probabilities are based on the Hadley Centre Ensembles.

Industrialization and urbanization are hard-wired into the development process; hence restricting these processes, and the attendant expansion of energy sources that they require, is not an option. Holding down emissions in developing countries will require not only a massive injection of renewable sources into the energy mix, along with improved energy efficiency, but also prevention of deforestation, changes in land-use planning, and reorganization of transport and water management. All of this entails major costs—costs that explain developing countries' objections to externally imposed emission commitments. Moving the climate agenda forward demands an integrated approach, a climate-inclusive developmental approach. Treating climate and development separately, as has largely been the case in the past, no longer provides a tenable framework for progress on either front.

Interrelated threats

Climate change and development are closely interconnected, with cumulative feedbacks and reactions in both directions, particularly through the production and use of energy. Economists, as suggested earlier, have a poor record when it comes to analysing these kinds of feedbacks and cumulative linkages. Policymakers, meanwhile, often appear predisposed to underestimate both the scale of the threats they are facing and the cost of removing those threats.

More recently, however, there have been growing signs of recognition of the urgency of the situation, and awareness that the international community faces a series of interrelated threats which can no longer be effectively tackled in isolation. A climate crunch, an energy crunch, a food crunch, and, most dramatically in recent years, a credit crunch have all exposed the danger of leaving risk management in the hands of the self-regulating forces of the marketplace.

Since 2008, policymakers in high-income countries have been struggling to deal with the interconnected shocks of a housing crisis, excessive energy consumption and financial collapse which have rippled and intensified throughout an increasingly fragile global economy. However, the challenges posed by the linkages among climate change, economic insecurity and political conflict are even greater for developing countries, and the consequences are likely to reach well beyond their own borders.

Adaptation without mitigation would ultimately be an ineffective response for developing nations, and the failure to deal with these interrelated threats would almost certainly have widespread and damaging consequences. It is increasingly clear that the climate and development problems will be solved together, and soon—or not at all.

A New Deal?

Parallels have often been drawn between the climate challenge and the mid-twentieth century experience of overcoming global economic crisis, defeating fascism and rebuilding ravaged economies. Al Gore, the former United States vice president, has suggested a Marshall Plan of sorts to tackle global warming (see chap. VI). Since the sharp downturn in the global economy starting in 2008, however, the emphasis has shifted to trying to develop a global new deal that can respond to the economic and climate threats simultaneously.

Historical analogies always need to be treated with a degree of caution, but as noted in chap. IV, the original New Deal, the United States' policy response to the 1930s depression, did address a series of interrelated threats through the kind of expanded and transformative policy agenda that is needed today. The scale of the response is also worth recalling: The New Deal committed 3 per cent of gross domestic product (GDP) to domestic programmes each year between 1933 and 1939, and much more was added in military spending to counter the threat of fascism during World War II. After the fighting stopped, the United States, through the Marshall Plan, committed almost 1 per cent of its GDP each year for five years to rebuild Europe. This was a massive investment over a 20-year period.

Economists have suggested that a smaller effort will be needed to respond to climate change. That seems optimistic. As Nicholas Stern indicates, the multi-decade strategy that is needed to keep climate risk manageable will involve long-term planning and massive investment and will require the kind of leadership and cooperation that helped defeat fascism and rebuild shattered economies.[11] Time is also of the essence; the sooner we start investing in a transition to low-emissions development, the lower the costs and the greater the chances of success will be.

No country left behind

Gauging the costs of climate change is difficult and requires reliance on modelling assumptions and scenarios. Damage functions are difficult to specify, externalities are difficult to price, and costs rise with the ambitiousness of the targets. Using standard economic models, the IPCC's *Fourth Assessment Report* estimated that the damage inflicted by climate change would entail, on average, a loss of 1 to 5 per cent of global GDP. However, the IPCC also noted that globally aggregated figures are likely to underestimate the damage costs, because they cannot include many "non-quantifiable impacts".[12]

Climate damages to developing countries are already perceptible. Indeed, one recent study estimated that for every 1° C rise in average global temperatures, annual average growth in poor countries drops by 1.1 percentage points, but there is no drop in rich countries.[13] Table I.5 presents two sets of estimates, from one of the widely used economic models, of the damages under a "business-as-usual" (BAU) scenario. Under either scenario—using either average or much worse than average assumptions—the damage to developing regions (measured as a percentage of projected GDP in 2100) is more than double that for OECD countries excluding the United States, and more than five times the damage to the United States.

Climate change is also multiplying vulnerabilities in developing countries by heightening livelihood risks and weakening adaptive capacities. More

Table I.5
Business-as-usual damages in 2100: The PAGE2002 model

	Annual damages as percentage of GDP in 2100			
Region	*Economic*	*Non-economic*	*Catastrophic*	*Total*
A. Mean damages in 2100: 'no adaptation' scenario				
United States	0.3	0.4	0.1	0.8
Other OECD	0.7	1.0	0.2	1.9
Rest of the world	1.6	2.3	0.4	4.3
World total	***1.2***	***1.8***	***0.3***	***3.4***
B. 83rd percentile damages in 2100, no adaptation, increased catastrophic risk and increased damages scenario				
United States	0.6	0.9	1.2	2.6
Other OECD	1.4	2.0	3.1	6.2
Rest of the world	3.2	4.5	6.3	13.5
World total	***2.5***	***3.6***	***4.8***	***10.8***

Source: Ackerman et al., 2009.

Notes: The results are based on a Monte Carlo analysis of 5000 runs of the PAGE2002 model. Both scenarios "turn off" the default PAGE assumptions about low-cost adaptation. Scenario B also modifies the probability distributions for catastrophic risk and for the relationship between temperature and non-catastrophic damages. See source for details. As the impacts are closely but not perfectly correlated, the 83rd percentile of total damages is slightly less than the sum of the 83rd percentiles of the individual damage categories.

than one-third of the world's population lives within 100 kilometres of a shoreline, facing threats from sea-level rise and storms; one in 15 people lives in coastal areas at elevations below 5 metres, where sea-level and storm surge risks are extreme.[14] In other areas, extended periods of drought have created a flow of environmental refugees and conflict with neighbouring countries. Similarly, tropical diseases are expected to become more common in areas with increased incidence of heat waves, while the prevalence of water-borne diseases is likely to rise in areas with an increased incidence of floods (see chap. III).

Populations that are already vulnerable because of low levels of economic and human development will be hit hardest by the growing threats from climate change. Poorer countries and communities with poor health care, lack of infrastructure, weakly diversified economies, missing institutions and soft governance structures may be exposed not just to potentially catastrophic large-scale disasters, but also to a more permanent state of economic stress as a result of climate impacts. That increased vulnerability, in turn, will deepen inequalities, with the least developed countries and small island States most affected.

In this context, adaptation is essentially a development challenge. It will require significant investments, not only to climate-proof existing projects and ensure effective responses to natural disasters, but also to diversify economic activity and address existing vulnerabilities. In many cases, adaptation will also dovetail with mitigation, which is equally important. For example, energy conservation measures could be classified under both mitigation and adaptation. Chap. III develops these arguments.

Common but differentiated mitigation challenges

Accumulating scientific evidence indicates that to prevent dangerous interference with the climate, atmospheric carbon concentrations should be stabilized at levels that keep temperatures from rising by more than 2° C above pre-industrial levels. As noted before, this would require reducing emissions by 50–60 per cent below 1990 levels by 2050, and even then, there is a serious risk. The longer we wait to start, the more rapidly we will have to reduce emissions later, and the greater the costs will be. Advanced countries should take the lead (see chap. II), both to reduce their own emissions, and to support the efforts of developing countries to establish a viable low-carbon development pathway.

Researchers have used both case-study evidence and modelling exercises to better understand the mitigation costs involved. Using the former approach, McKinsey & Company has developed a ranking of mitigation steps in order of increasing marginal costs (see chap. II). Others have identified "wedges" of alternate technologies, such as automobile efficiency improvements, increased reliance on renewable energy, or carbon capture and storage, that could displace a certain amount of emissions each year.[15] The alternative has been to use integrated assessment-type models to estimate mitigation costs, but the two approaches are not mutually exclusive.

While the absolute values of required investment can appear quite high, the costs of inaction are even higher. It is also clear that the lower the stabilization level chosen, the safer the future, but the higher the initial investment costs. As noted above, very broadly, even an annual cost as high as 2 or 3 per cent of global GDP is small in comparison with the potential damage from following business-as-usual pathways. Thus the benefit-to-cost comparison is strongly in favour of rapid action to mitigate climate change.

Defining low-emissions, high-growth pathways

The shift to a low-carbon economy will pose different challenges for countries at different levels of development. For advanced countries, the emphasis may be on increasing employment and energy security, the core of the so-called "green jobs" agenda. For many developing countries, diversifying economic activity away from the primary sector and low value-added manufacturing remains an essential policy goal, combined with efforts to eradicate poverty and ensure a more balanced integration into the global economy.

Incremental change or a big push?

There are few historical precedents for the kind of transitions needed. One approach is to create incentives for private businesses to shift gradually from high-emitting activities and make investments in new high-risk, high-return, climate-friendly technologies. To create the right investment climate, it has been argued, there also has to be a good governance agenda that establishes a price for carbon, guarantees strong intellectual property rights and removes distortionary subsidies for high-emitting activities.

Alternatively, the transition could be driven by a series of large and long-lasting investments in energy infrastructure, land-use changes, transportation, etc.—which would have to be undertaken simultaneously to have a significant impact on climate change. This is the approach adopted by this book and one also advocated by leading specialists, such as experts of the United Nations Committee for Development Policy (CDP).[16] Price incentives by themselves are unlikely to trigger or sustain the required investments. Rather, a "big push" is needed to launch a successful low-emissions development pathway. This revives long-standing questions about the challenges facing poor countries in mobilizing investment resources and the relative roles of the public and private sectors. It also draws attention to the scale of the financing challenge involved; these issues will be discussed in greater depth in chaps. II, IV and VI.

Does technology hold the key?

The dual challenge of meeting development goals, including rapid industrialization, while controlling emissions and reducing carbon dependence will require the creation of new, scalable and powerful technologies in the next 10 to 20 years—technologies that transform not only how energy is produced, distributed and used, but also how we help vulnerable countries adapt to rising temperatures.

While there is broad agreement that technology will play a central role in meeting this dual challenge, there is less of a consensus on how to build technological know-how and capacity, particularly given the significant gaps between rich and poor countries. For some, stronger protection of intellectual property rights, both to encourage local innovators and to attract foreign direct investment, is key to leapfrogging over old technologies onto a cleaner technological highway. Others not only doubt the efficacy of such mechanisms in generating the required level of innovative effort, but also view them as a source of potentially significant obstacles for developing countries (see chap. V).

Experience indicates that, in many successful past instances of technology development, government support has extended beyond research and development, for example, through government procurement and loan guarantees for construction and equipment investments. Yet some question the wisdom of this approach, pointing to the difficulty of second-guessing the market and "picking winners". Others worry about the high costs of learning and uncertainty that come with experimentation. Such initiatives, whether undertaken by the private or the public sector, constitute grounds for socializing the risks involved. These issues are discussed in greater detail in chaps. IV and V.

An expanded public policy agenda

If climate is, indeed, a global public good, then stopping free-riding, strengthening property rights, and ensuring good collective governance would seem to be the main resulting challenges. But that may be too narrow a framing, in part because the allocation of the atmosphere's limited remaining capacity to absorb emissions, as well as the distribution of the costs of abatement, are contentious, high-stakes issues. Moreover, achieving fast growth in developing countries and high employment in advanced countries along low-emissions pathways will entail complex decisions and trade-offs regarding consumption, settlement, transportation and urbanization. The climate challenge is also difficult to separate from food and energy security and global health issues.

Many disagreements on climate policy stem from diverging views on how best to meet these challenges: should there be a gradual shift away from business-as-usual scenarios or a transformative change? A central question concerns the relative roles of the private and the public sectors in making the needed investments. One way in which governments can act is by instituting a carbon price, through a carbon tax or a cap-and-trade policy, along with strong regulations. Much of the discussion of the climate challenge in developed countries is focused on the relative efficacy of alternative ways of establishing a carbon price. The policy mix in developing countries is likely to be different, with a much larger role for public investment and targeted industrial policies. At every level of development, all policy instruments, ranging from price incentives, taxes, subsidies, and regulation, to fiscal, monetary and financial measures, should be considered.

A low-emissions, high-growth scenario

To assess the implications of various climate and development investment scenarios, a simulation was run with the Global Policy Model (GPM) developed in the United Nations Department of Economic and Social Affairs (UN/DESA).[17] The model was created to investigate the spillover effects of macroeconomic policy scenarios in an interdependent world economy, but it also spells out energy production and demand for country groups and an international market (a pool) which sets the equilibrium price. The model does not specify carbon emissions linked to economic activity; therefore, inferences regarding climate change scenarios are drawn from trends in energy efficiency and energy use.

The business-as-usual (BAU) scenario used as the basis for the analysis assumes that after 2010, the world economy has recovered from the financial crisis and returns to the past pattern of growth. This entails a continuation of current trends in (high-emissions) energy intensity, and in economic inequality. The implication is that, in the BAU scenario, the world resumes growth on a path deemed unsustainable from both a development and an environment perspective.

The alternative, low-emissions, high-growth scenario, built as a policy-driven departure from BAU, requires international policy coordination. Three types of policy adjustment are considered:

- Countries worldwide are assumed to increase public spending levels by 1 to 5 per cent of GDP, with developed countries in the lower end of the range and developing countries in the upper end. The investment push is expected to trigger faster economic growth, promote energy efficiency, and help increase the supply of primary commodities and food at a rate consistent with the growth of world income.
- The investment push and international agreements contribute to reducing high-emissions energy demand (reflecting, for instance, a cap-and-trade mechanism), yielding lower emissions and greater energy efficiency. Such improvements in energy efficiency are consistent with the investment patterns discussed below.
- Economic resilience of developing countries is strengthened by providing those countries, especially the poorest, with full, duty-free market access to developed-country markets, leading to greater economic diversification.

Energy efficiency and energy diversification

To assess the implications of changing course, the model looks at the impact of increased public investments in infrastructure, diversification of economic activity and energy provision by governments worldwide. As discussed further in chap. IV, after possible financial "crowding-out" mechanisms are accounted for, such public spending is found on balance to "crowd in" (i.e., stimulate an increase in) private investment.

The assumption that growth in investment has the potential to boost energy efficiency is based on empirical evidence for a number of countries (see table I.6). Accordingly, the policy scenario assumes a boost in public investment, stimulating growth in total investments. Table I.7 summarizes the outcomes at the end of the simulation period in 2030.

Table I.6
Energy use and total investment, selected countries: 20-year averages, 1990

	Efficiency: change in energy use per unit of output (percentage per year)	*Rate of growth of total investment (percentage per year)*	*Elasticity: ratio of efficiency gain to investment growth*
Switzerland	-1.18	2.10	0.6
Finland	-2.03	4.31	0.5
France	-3.21	3.30	1.0
Sweden	-5.79	2.59	2.2
Japan	-1.98	4.15	0.5
United States	-2.94	3.02	1.0

Source: United Nations, *Energy Statistics Yearbook and National Accounts Statistics*, various years.

Table I.7
Energy use and total investment: 20-year averages, projected for 2030

	Efficiency: change in energy use per unit of output (percentage per year)	*Rate of growth of total investment (percentage per year)*	*Elasticity: ratio of efficiency gain to investment growth*
Developed countries	**-5.20**	**2.90**	**1.80**
Japan	-5.00	3.75	1.30
Europe	-4.80	2.92	1.60
United States	-5.40	2.54	2.10
Developing countries	**-5.80**	**6.80**	**0.90**
China	-6.40	6.45	1.00
Least developed countries	**-6.65**	**9.90**	**0.70**

Source: United Nations, Department of Economic and Social Affairs, Global Policy Model simulations. See text for details.

Such results may seem implausible at first sight, but they are within the range of recent experience. Based on the elasticities (ratio of efficiency gains to investment growth rates) shown in tables I.6 and I.7, the model simulation assumes that developing countries can achieve results comparable to those in a number of countries in the recent past. Developed countries would be achieving high efficiency improvements for the amount of investment, almost as high as in the best of the cases in table I.6.

The effects of these efficiency improvements on fossil fuel use and CO_2 emissions cannot be precisely established with this model. The model projects world economic growth of about 5 per cent per year, and, under the policy scenario, a reduction in energy use per unit of output of 6 per cent per year. Thus global energy use, measured in physical units, declines at an annual rate of about 1 per cent.

While of great importance, this gradual decline in energy use alone is not sufficient to reduce global greenhouse gas emissions by 50–60 per cent by 2050. Energy efficiency improvements will need to be complemented by massive investments in renewable low-emissions energy sources, as assumed in the model simulations, leading over time to a drastic change in the composition of energy sources.

How rapidly can non-fossil energy use expand? In a 2007 study, UN/DESA and the International Atomic Energy Agency found that, between 1980 and 2000, biofuels and hydroelectricity supplied about 40 per cent of Brazil's total demand for energy, and grew at an average rate of 2.25 per cent per annum.[18] Significantly better records have been obtained in France through its shift to nuclear energy. Biofuels and nuclear energy are not, of course, problem-free alternatives. However, other sources, such as wind, solar and hydroelectric power, are valid options and are likely to become far more efficient as technologies advance. Recent surges in installed wind and solar power capacity in many countries are the natural result of growing experience, increasing efficiency, and plummeting costs.[19]

Financing or access to markets?

There is no doubt that a low-emissions, high-growth strategy will carry high initial costs for both developed and developing economies. Developed countries are better prepared to carry it out, because they have the financial and technological resources; but even if they achieve the kind of targets proposed above, success in developed countries alone will not suffice to meet global climate goals.

Thus a plan is needed to supply the resources needed to launch such a strategy in the developing world. The initial investment push, as discussed in chap. VI, will inevitably require financial support from developed to developing countries, particularly the least developed countries. For the longer term, the goal is to make developing countries self-sufficient, creating a sustainable system of financing for their climate and development investment needs. The Global Policy Model scenario presented here assumes concerted action by policymakers, particularly in industrialized economies, to improve developing countries' access to their markets for manufactures and services. If this is accompanied by an international accord that encourages steady-state growth of production of food and primary materials and thus stable terms of trade (as is the case for agricultural prices in the European Union and elsewhere), their rapid expansion will benefit developing and developed countries alike.

Once there is a plan to increase developing countries' market share in manufactures and services, their need for external resources will diminish sharply. Furthermore, in the absence of an external debt burden, a combination of stable commodities prices and sustained income growth in both the developing and the developed world will contribute to reducing the fluctuations in domestic prices, interest rates, exchange rates, etc., helping to avert the stop-go adjustment-stabilization processes which have been so damaging for long-term development over the last few decades.

Assessing the simulation results

This modelling exercise assessed the feasibility of a low-emissions, high-growth path from an economic point of view, finding that it clearly is feasible. It reduces absolute energy consumption despite sustained rates of global economic growth. It also yields significantly higher rates of growth in the developing world, and it allows developed countries to grow faster than under business as usual. The critical factor is public investment-led expansion on a significant, but not historically unprecedented, scale. In terms of income per capita, this scenario yields an improvement for all blocs, most dramatically for poorer countries. Finally, it contributes to export diversification, stable terms of trade, and a smooth reduction of external imbalances that have proved to be unsustainable.

The potential problems with this scenario are not attributable to the underlying economic principles of the model simulation, but rather to the political processes that are required for such a big push to take place. Without serious international policy coordination, this scenario cannot work. Hopefully the seriousness of the environmental challenge will impel policymakers to embrace a common goal such as this.

Conclusion: managing crises

John Maynard Keynes famously remarked that "in the long run we are all dead." Keynes was responding to policymakers in the early 1920s who were postponing urgently needed action to counter immediate economic hardships, in the belief that market forces would (in the long run) bring the desired recovery. Similar thinking has informed much of economic policymaking during the past three decades. His quip takes on a much more ominous meaning, however, in light of the combined threats to our economic and environmental security.

Price shocks during 2008 and again during 2010–2011 in food, fuel and financial markets laid bare the world economy's shaky foundations—excessive debt, unregulated capital flows and rampant speculation. The cost in terms of declining asset values and government bailouts of collapsed financial institutions was staggering, followed by more widespread damage from a multi-year slump in the economies of advanced, emerging and least developed countries alike.

Looking ahead, policy responses must be designed not just to help meet the short-term goals of creating jobs and securing homes, but also to achieve longer-term security goals, including a stable climate. Turning the page on "casino capitalism" and establishing truly sustainable low-emissions alternatives will require policymakers to draw some hard lessons from recent experience. They must realize, for example, that markets in general—not only financial markets—do not regulate themselves, but depend on an array of institutions, rules, regulations and norms to correct coordination failures, moderate their more destructive impulses, and manage the tensions these impulses can generate.

The shift to a low-emissions, high-growth development path requires a clear break from recent policy approaches, as well as a long-term commitment to doing what it takes to create jobs in advanced countries and support growth in poorer countries. It will involve smarter incentives, stronger regulations and, above all, significant investments, including in the public sector.

The economic crisis that began in 2008 serves as a reminder that financial institutions need to get back into the business of securing people's savings and of building stable networks and levels of trust that can support more socially productive investment opportunities. These policy challenges are of long standing in many developing countries, where financial markets have repeatedly failed to build long-term commitments. Adding in the climate challenge only reinforces the urgency of reforming the financial system, given the scale of resources that will have to be mobilized over the coming decades and the trade-offs that will have to be made if economies are to secure a low-emissions future.

Market forces have an important role to play, but real leadership will have to build upon a strong public policy agenda and a revitalized social contract—at both the national and international levels. Markets are prone to generate incomplete or incorrect information, giving rise to perverse behaviour (ranging from moral hazard and free-riding to outright fraud) and undesirable outcomes (excessive leverage, the proliferation of toxic products, hidden accounting practices). Both the strengths and the weaknesses of price incentives need to be kept firmly in mind as market-based solutions are extended to meet the climate challenge.

The immediate response to the financial crisis provided a reminder that governments are the only agents capable of mobilizing the massive financial and political resources needed to confront large systemic threats. It also showed that policymakers can act with real urgency when faced with such threats. This is encouraging from both the development and the climate angles, given that both challenges involve large resource commitments over the long term, and at both the national and the global levels. Unfortunately, the subsequent years have shown that it is not easy to maintain this agenda after the moment of greatest crisis subsides. To meet the climate and development challenges, we must not only surmount traditional market failures that occur as a result of externalities and free-riding, but also deal with systemic threats, manage large-scale adjustments in economic activity—and stay with this agenda for the long haul. The only sensible response is to mix market solutions with other mechanisms, including regulations and public investment, and to commit ourselves to active, ongoing involvement.

Notes

1 United Nations Environment Programme, 2010. *The Emissions Gap Report: Are the Copenhagen Accord Pledges Sufficient to Limit Global Warming to 2°C or 1.5°C?* http://www.unep.org/publications/ebooks/emissionsgapreport/.

2 United Nations, 2011. *World Population Prospects: The 2010 Revision*. UN/DESA Population Division. http://esa.un.org/unpd/wpp/.

3 Food and Agriculture Organization of the United Nations, 2010. *Global Forest Resources Assessment 2010*. FAO Forestry Department, Rome, Italy. http://www.fao.org/forestry/fra/en/.

4 World Resources Institute, 2010. 'Climate Analysis Indicators Tool.' *CAIT 8.0*. http://cait.wri.org/.

5 Cazenave, A. and Llovel, W., 2010. 'Contemporary Sea Level Rise.' *Annual Review of Marine Science*, 2(1). 145–73. doi:10.1146/annurev-marine–120308–081105.

6 Richardson, K., Steffen, W., Schellnhuber, H. J., Alcamo, J., Barker, T., Kammen, D. M., Leemans, R., Liverman, D., Munasinghe, M., Osman-Elasha, B., Stern, N. and Wæver, O., 2009. *Synthesis Report: Climate Change: Global Risks, Challenges*

& Decisions. Synthesis report for Climate Congress held March 3–5, 2009, in Copenhagen. Key Message 1. http://climatecongress.ku.dk/pdf/synthesisreport.

7 Stern, N., 2006. *The Economics of Climate Change: The Stern Review*. Cambridge University Press, Cambridge, UK. http://www.hm-treasury.gov.uk/stern_review_report.htm.

8 Kindleberger, Charles, 1986. 'International public goods without international government.' *American Economic Review*, 76(1), 1–13.

9 For a more in-depth look at this issue, see Stanton, E. A., Ackerman, F. and Kartha, S., 2009. 'Inside the Integrated Assessment Models: Four Issues in Climate Economics.' *Climate and Development*, 1(2). 166–84.

10 Helm, D., 2008. 'Climate-change policy: Why has so little been achieved?' *Oxford Review of Economic Policy*, 24(2). 211–38. doi:10.1093/oxrep/grn014.

11 Stern, N., 2009. *A Blueprint for a Safer Planet*. Random House, New York. 12–13.

12 The IPCC also reported that peer-reviewed estimates of the "social cost of carbon" (net economic costs of damages from climate change aggregated across the globe and discounted to the present) for 2005 had an average value of $12 per ton of CO_2, but the range of 100 estimates was large, from -$3/t$CO_2$ to $95/t$CO_2$.

13 Dell, M., Jones, B. F. and Olken, B. A., 2008. *Climate Change and Economic Growth: Evidence from the Last Half Century*. NBER Working Paper No. 14132. National Bureau of Economic Research, Cambridge, MA. http://www.nber.org/papers/w14132.

14 Population within 100 km of the coast is taken from United Nations Environment Programme and GRID-Arendal, 2007. 'Coastal population and altered land cover in coastal zones (100 km of coastline).' *UNEP/GRIP-Arendal Maps and Graphics Library*. http://maps.grida.no/go/graphic/coastal-population-and-altered-land-cover-in-coastal-zones-100-km-of-coastline. Population at 5 metres or less elevation is taken from Center for International Earth Science Information Network, Columbia University, 2007. 'National Aggregates of Geospatial Data: Population, Landscape and Climate Estimates, v.2 (PLACE II).' http://sedac.ciesin.columbia.edu/place/.

15 See, for example, Grubb, M., 2004. 'Technology innovation and climate change policy: An overview of issues and options. *Keio Economic Studies* (Japan), 41(2). 103–132.

16 United Nations, 2007. *The International Development Agenda and the Climate Change Challenge*, Committee for Development Policy, Policy Note. New York: United Nations. http://www.un.org/en/development/desa/policy/cdp/cdp_publications/climate_07.pdf

17 For a more detailed description of this model, see United Nations, 2009. *World Economic and Social Survey 2009: Promoting Development, Saving the Planet*. E/2009/50/Rev.1. New York: United Nations. http://www.un.org/esa/policy/wess/wess2009files/wess09/wess2009.pdf

18 United Nations and International Atomic Energy Agency, 2007. *Energy indicators for sustainable development: Country studies on Brazil, Cuba, Lithuania, Mexico, Russian Federation, Slovakia, and Thailand*. New York. http://www.un.org/esa/sustdev/publications/energy_indicators/full_report.pdf

19 The feasibility of accelerated diffusion of clean energy technologies and of technologies for low-emission agricultural farming has been analyzed in greater depth in United Nations, 2011. *World Economic and Social Survey 2011: The Great Technological Transformation*, New York. http://www.un.org/en/development/desa/policy/wess/wess_current/2011wess.pdf

Chapter II
Climate mitigation and the energy challenge

Introduction

A maximum temperature increase of 2° C above pre-industrial levels is the consensus science-based target for prevention of dangerous, perhaps irreversible changes in the earth's climate; this can only be achieved by very rapid reduction in greenhouse gas emissions. At the same time, developing countries need to achieve a sustained annual growth rate of at least 6 per cent to "catch up," or close the income gap with developed countries within a reasonable length of time. Those two broad objectives frame the mitigation and development challenge facing policymakers at the national and international levels. This chapter looks at mitigation options that would allow for both climate stabilization and rapid economic growth in developing countries.

The 2° C target, according to recent analyses, requires reducing global emissions by 50–60 per cent from 1990 levels by 2050, or about 3 per cent annually, with a peak no later than 2020. This is no small undertaking and requires significant economic adjustments in both developed and developing countries. There are significant win-win options (where energy savings pay for mitigation investments in a short period of time), particularly in energy efficiency, but there are not nearly enough of these costless options to meet stabilization targets. Large-scale, upfront investment in electricity production will be needed, including new sources of renewable energy, along with related investments in transportation and construction.

What is required is a gale of "creative destruction" driven by massive investments and innovative technologies. This is not inevitable but will require dedicated and strategic policy action at all levels. As Nicholas Stern has warned, there is a danger that if we delay such action, current and near-term investment projects will lock in older technologies for decades, leading to a ratcheting up of the stock of emissions to dangerous levels, and requiring much-costlier economic and social adjustments in the future.

Despite the accumulating scientific knowledge and growing public awareness of climate change, effective mitigation action has lagged in developed countries, primarily due to a persistent disconnect between environmental and economic objectives. For developing countries, meanwhile, effective mitigation requires not only a change in global and national focus on climate and development policy, but also strategic thinking to integrate mitigation efforts with poverty reduction, rural development, energy access, industrial expansion and infrastructure provision.

The energy sector, broadly defined, accounts for 64 per cent of global emissions (table I.1 in chap. I), or 28.4 gigatons (Gt) of carbon dioxide equivalent (CO_2e) in 2005. Within this category, electricity and heat produce the largest share of emissions, 12.4 Gt CO_2e in 2005; transportation produces about one-fifth of energy-related emissions, as does manufacturing and construction. Without significant changes in energy production and consumption, it will not be possible to meet stabilization targets.

The next largest source of emissions is agriculture (particularly rice paddies and livestock), which produced 6.1 Gt of CO_2e in 2005, or nearly 14 per cent of global emissions, surpassing deforestation, which accounted for about 12 per cent. Deforestation does remain a major concern, especially in developing countries: in 2005, land use and deforestation accounted for 21 per cent of emissions in non-Annex I countries and 48 per cent in least-developed countries.[1]

This chapter focuses on energy in particular because it is the pivotal issue at the interface of climate and development. This is not to say that mitigation options in other sectors are unimportant—but unless the energy challenge is addressed, developing countries will not be able to mitigate their emissions or transform their economies to meet climate and development goals.

Stabilization scenarios and mitigation options

The more than 200-fold growth in global CO_2 emissions between 1750 and the present has resulted in a dramatic increase in atmospheric concentrations, from about 280 parts per million (ppm) to more than 390 ppm of CO_2. The increase in CO_2 emissions is the most important contributor to global warming, but other greenhouse gas emissions have also increased rapidly, including methane, several other gases, and a fundamentally new, man-made addition to the atmosphere: chlorofluorocarbons (CFCs). Further complicating the picture is the fact that sulphur and particulate-matter emissions, posing other health and environmental hazards, are caused by the combustion of coal and other fossil fuels. Yet despite the importance of other greenhouse gases—and, in many cases, the low cost of reducing their emissions—the heart of the problem is still the reduction in CO_2 emissions.

The climate challenge demands fundamental changes in the global energy system, land-use patterns and human behaviour. Managing those changes will require an integrated policy framework to move from current emissions-intensive patterns of wealth creation to a future low-emissions global economy. Timely, widespread technological improvements will be of the utmost importance, including both new technologies, and induced changes to existing technologies.

The energy intensity of the global economy will have to be cut to one-half or even one-third of historical levels. All stabilization scenarios call for a huge share of emissions reductions, in the range of 60–80 per cent, from changes in energy systems. The solutions will vary across regions, with different shares of renewable energy, nuclear energy, carbon capture and sequestration (CCS), biomass and hydrogen and other advanced energy carriers. Energy efficiency will also play a crucial role, though as noted in chap. I, it can by no means fully solve the problem.

Achieving low stabilization levels will require early (upfront) large-scale investments and substantially more rapid diffusion and commercialization of advanced low-emissions technologies. Such investments will need to be made worldwide on the required scale, implying that effective technology and resource transfers will need to be made to countries lacking those means (see chaps. V and VI).

Currently, there are several options for curbing emissions without jeopardizing economic growth, especially in developing countries. These include a switch to renewable energy technologies (most notably wind and solar power), the enhancement of terrestrial sinks through afforestation in conjunction with sustainable biomass use, and investment in energy efficiency solutions. Another major option, the use of CCS technologies at fossil fuel plants, is under development but not yet ready for widespread adoption.

The greenhouse gas abatement cost curve developed by McKinsey & Company provides a useful quantitative estimate of both the costs and the actions needed to achieve such reductions (figure II.1). The curve ranks technologies and industrial processes according to the net costs of avoiding a ton of CO_2 emissions, taking into account both the capital costs and the operating costs of low-emissions technologies. Figure II.1 starts with opportunities for negative cost (or win-win) emissions reductions where the upfront capital costs are more than offset by future energy savings. Most of these savings are achieved through improved energy efficiency. The curve then progresses through reduction opportunities with positive, although often very

Figure II.1
Global GHG abatement cost curve beyond business-as-usual, 2030

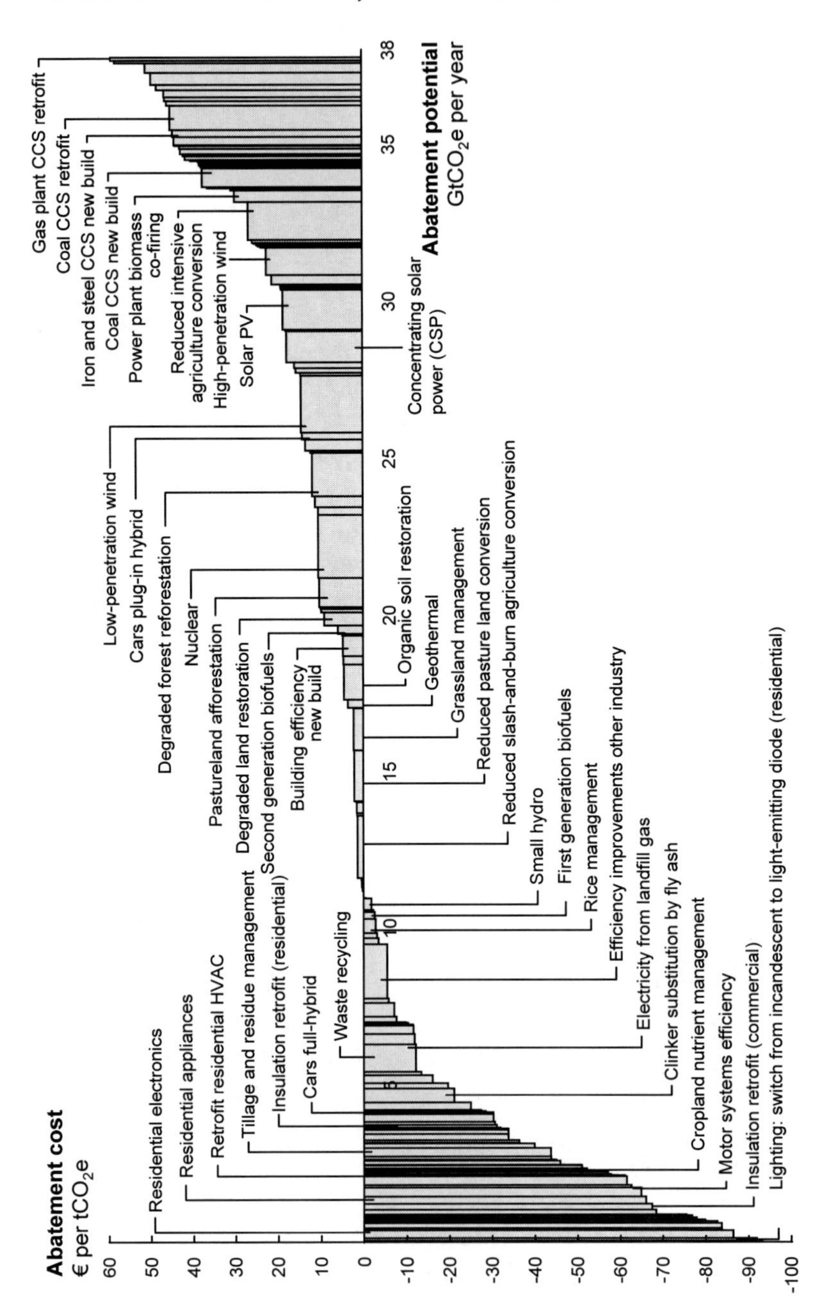

Source: McKinsey & Company, 2009.

low, net costs. Technical abatement opportunities up to a cost of €60 per ton of CO_2e include energy efficiency, low-emissions energy supply, and forestry and agriculture (figure II.2). These options generate a total abatement of 38 Gt CO_2e per year in 2030 relative to annual business-as-usual emissions of 70 Gt CO_2e. Abatement opportunities in these three categories are spread across many sectors of the economy, with approximate figures of 29 per cent for the energy supply sectors (electricity, petroleum and gas); 16 per cent in the industrial sector; 22 per cent in transport, buildings and waste; and 33 per cent in land-use sectors (forestry and agriculture). In all, developing countries have 70 per cent of the reduction opportunities, while developed countries have 30 per cent.

The central feature of these options is that they assume an immediate start up—a delay of 10 years would almost certainly mean missing the 2° C target. Many developing countries are already taking steps on mitigation. However, more action will be required. The policy challenge is to ensure that such action supports, rather than obstructs, the achievement of development goals.

Figure II.2
Major categories of abatement opportunities

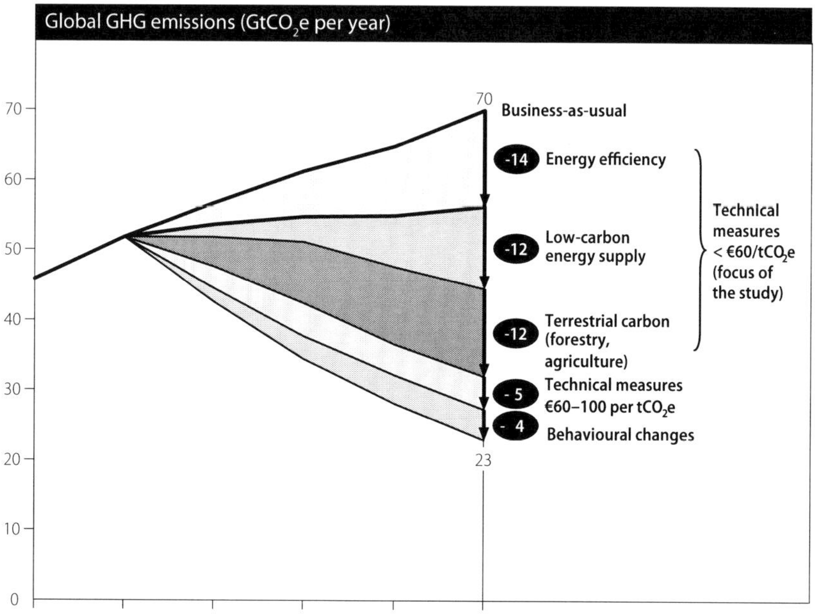

Source: McKinsey & Company, 2009.

Energy and economic development

The evolution of the energy system

In 1800, the world's population was approximately one billion, representing a little more than a fourfold increase over the population in AD 1. With the spread of industrialization, things changed radically. In the next two centuries, global population grew sevenfold, to 7 billion (see table II.1; note that it has since grown by almost another billion)—corresponding to an annual growth rate of close to 1 per cent, and a doubling of population every 80 years. This sustained population growth has been a result of drastic decreases in mortality and increases in longevity. Improved water quality, diet, sanitary conditions and medicine have all contributed, and all are correlated with increased availability of energy resources.

Gross world product increased much faster than population: production grew more than 100-fold during the last two centuries, corresponding to an average increase of more than 2 per cent per year and a doubling every 32 years. To a large extent, this was made possible by the replacement of human workers and draft animals with machines fuelled by fossil energy, and the resulting release of labour in high-productivity activities such as manufacturing.

Technological advances have gradually reduced the energy-intensity of the global economy, so that energy use increased "only" 41-fold over the same span of time. Most of that surge in energy use came from fossil energy sources, with the unavoidable by-product of rising CO_2 emissions. Additional technological advances, and the shift from coal to oil and gas,

Table II.1
Population, economic activity, energy use and emissions, 1800–2010

	1800	*About 2010*	*Ratio (2010/1800)*
Population (billions)	1.0	6.9	6.9
World gross product (trillions of 2005 US dollars)	0.5	50	100
Primary energy use (exajoules)	13	530	41
CO_2 Emissions (Gt CO_2)	1.1	29	26
Mobility (km/person/day)	0.04	40	1 000

Sources: Nakicenovic, 2009 for 1800 data. The sources for the data for 2010 or most recent are as follows: population (2010) from United Nations, World Population Prospects: The 2010 Revision, Population Division, Department of Economic and Social Affairs; world gross product (2010) from UN/DESA database for World Economic Situation and Prospects; primary energy use (2009) from United Nations, World Economic and Social Survey 2011: The Great Technological Transformation, New York, p. 21; CO2 emissions (2008) from the International Energy Agency (http://www.iea.org/stats/indicators.asp?COUNTRY_CODE=29); and the mobility indicator (2000) from Nakicenovic, 2009.

slowed the increase in energy-related CO_2 emissions to a mere 26-fold, to about 29 Gt of CO_2 in 2000. Thus there has been a long-term trend toward lower energy intensity, and toward lower carbon intensity—but these trends have only moderated, not stopped or reversed, the relentless effects of economic growth.

In 1800, the world still depended on traditional biomass (mostly wood and agricultural waste) as the main energy source for cooking, heating and manufacturing. Human physical labour and animals were the main sources of mechanical energy, with some, but much more humble, contributions from wind and hydraulic power. By 1850, coal already met some 20 per cent of global primary energy needs; the figure rose to almost 70 per cent by the 1920s (figure II.3). This shift may be characterized as the first energy transition. The coal age brought such transformative technologies as railways, steam power, mass production of steel, large-scale manufacturing, and the telegraph.

Around 1900, motor vehicles were introduced along with petrochemicals and other technologies that depend on petroleum. It took another 70 years for oil to replace coal as the dominant source of energy in the world. Today, the global energy system is much more complex, with many competing

Figure II.3
Global primary energy requirements since 1850

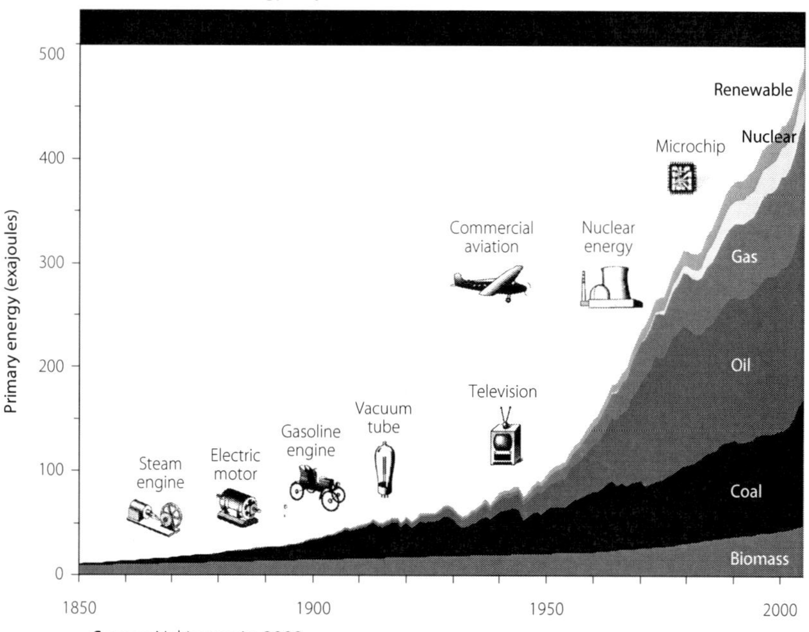

Source: Nakicenovic, 2009.

sources of energy and many high-quality and convenient energy carriers. Electricity is most often produced from coal, natural gas, or renewables; liquids (petroleum products, or ethanol) are primarily used in transportation; various forms of biomass are still widely used in the developing parts of the world (where one third of the global population still does not have reliable access to modern energy services). Taken together, fossil energy sources provide some 80 per cent of global energy needs, while wood, hydropower and nuclear energy provide almost all of the rest.

Energy and growth

Energy is the critical link between development and climate mitigation. Access to energy services, like income, is distributed very unequally, with a strong correlation between the two.

Development economists have long promoted significant investments in "social overhead capital" such as the provision of energy services, in part because of the direct welfare effects of the services provided, but also because of their potential to crowd in other productive investments. Returns on these investments are likely to be highest in the early development stages, when basic networks are still incomplete. In low-income countries, basic services such as water, irrigation and transport account for most infrastructure spending, while in middle-income countries, telecommunications, and especially electric power, become more important. Once the social overhead capital is in place, more targeted policy incentives can support further diversification and technological upgrading. Indeed, a virtuous circle of strong investment, rising productivity, falling costs, and expanding incomes and markets, leading to further investments and increases in productivity, exhibits the mix of cumulative supply- and demand-side impulses essential to sustained development. Large public investments in areas such as energy services can play a catalytic role in this.

A major goal of any big public investment programme is to increase the return on private investments in new technologies by creating rents and market opportunities for the private sector (see chap. IV). As Albert Hirschman recognized long ago, the success of such a push is not measured solely by the speed with which cost advantages in the targeted sectors are realized, but also by the links those sectors establish backwards to suppliers and forward to new activities and markets that use their products, whose expansion could trigger new investment opportunities.[2] Indeed, a high density of such linkages among modern or modernizing sectors is an essential dimension of development, and a key to sustained increases in incomes. Hirschman associated those linkages mainly with large-scale industrial investment, but he also recognized that the power sector had very strong linkage potential that could trigger cumulative development prospects.

The importance of electrification to rural development is well known. Major investments in rural electrification projects, mainly grid extensions, have been integral to successful growth experiences. In rapidly developing agricultural regions, electricity helps raise the productivity of local agriculture, industry and commerce by supplying motive power, refrigeration, lighting and process heating. Increased earnings from those activities, in turn, lead to greater household demand for electricity. Cheaper and better lighting can make education more productive, help build human capital, and raise output by extending the length of the workday and making working conditions more predictable.

Achieving convergent economic growth and energy consumption

Globally, about 31 million tons of oil equivalent are consumed daily as primary energy, equivalent to 55 kilowatt hours (kWh) per person per day. This consumption is distributed very unequally (see table II.2). In high-income countries, roughly coinciding with the membership of the Organization for Economic Cooperation and Development (OECD), average per capita consumption ranges from 100 to 300 kWh per day, divided roughly equally between household and commercial consumption. In the vast majority of developing countries, average per capita consumption is under 50 kWh per day. The exceptions are the countries of the Organization of the Petroleum Exporting Countries (OPEC), the newly industrialized countries and regions (Singapore, Republic of Korea, Hong Kong Special Administrative Region of China and Taiwan Province of China), and some emerging economies (such as South Africa). Most countries in sub-Saharan Africa and South Asia consume less than 20 kWh per capita per day. The differences are even wider in the consumption of electricity, the pre-eminent form of modern energy service, and the very symbol of modernity and affluence.

The threshold of 100 kWh per capita per day can be used as a convenient dividing line between energy poverty and energy sufficiency. Achieving this target worldwide would imply a significant expansion of energy infrastructure. Here is where the climate and energy agenda of developing countries begins to diverge from that of developed countries.

Table II.2
Per capita energy consumption, selected countries, 2008

	Population (millions)	*Primary energy (kWh per capita/day)*	*Electricity (kWh per capita/day)*
Canada	33.3	254.98	46.72
Germany	82.5	129.53	19.50
Japan	126.5	124.85	22.31
Sweden	9.2	171.06	40.66
United Kingdom	61.3	108.40	16.64
United States	305.0	238.59	37.33
Russian Federation	143.2	152.85	17.48
Brazil	191.5	41.34	6.13
Mexico	110.6	52.02	5.32
Venezuela, Bolivarian Republic of	28.1	72.80	8.39
Nigeria	150.7	23.51	0.35
South Africa	49.3	86.89	12.90
Egypt	78.3	28.77	4.07
Bangladesh	145.5	6.12	0.63
India	1,190.9	16.61	1.48
China	1,328.3	50.77	6.71
Hong Kong SAR	6.9	65.05	16.19
Indonesia	235.0	26.94	1.57
Korea, Republic of	47.7	151.49	24.70
Singapore	4.8	123.66	22.74
Thailand	68.3	50.03	5.62

Sources: UN/DESA, based on primary energy data (Total Primary Energy Supply in Mtoe per year) and electricity data (electricity consumption in TWh per year) from International Energy Agency; and population data from United Nations Population Division, DESA.

Developed countries have greater potential for energy conservation and energy efficiency, especially since most consume well over 100 kWh per capita per day; scaling down energy consumption could be consistent with the same or higher levels of income and well-being. In developing countries, in contrast, while energy efficiency is still important, it does not obviate the need for expansion of the energy infrastructure. Most countries will need to expand energy services to the threshold level of 100 kWh per day in order to meet their human development targets.

The second reason for divergence hinges on the question of affordability. Currently, the expansion in energy services in developing countries is hindered by the fact that the vast majority of the population is too poor to afford these services without some form of subsidy. Even populations with incomes of $3 per day would not be able to spend more than, say, $0.30 per person per day on energy-related expenditures (electricity, cooking, heating, transport). Even if energy is priced as low as, say, $0.05 per kWh, they would be able to afford only 6 kWh per day, comparable to the average in China or Brazil (see table II.2); the threshold of 100 kWh per day is well beyond their horizon.

This points to the need for a three-part agenda. At the aggregate level, it would make sense to set a minimum global target of 100 kWh per capita per day to overcome energy poverty. Second, energy efficiency measures should be instituted to ensure that this consumption level corresponds to the achievement of economic and human development targets. Third, at the most urgent level, there is a need to address "energy destitution", the lack of access to modern energy services.

The faster-growing developing countries have been able to follow this trajectory with reasonable success. However, the example of China shows the potential problems. Since 2000, China has more than doubled its energy consumption, to 2.26 billion tons of oil-equivalent in 2009, according to the International Energy Agency (IEA)[3]—and to a great extent, it has done so by exploiting coal, the cheapest energy source but also the worst in terms of global warming impact per unit of energy. China has also become a world leader in wind and solar power, but these technologies remain far costlier than coal; much of China's production of renewable energy technologies is for export, rather than for domestic use. For developing countries, relying heavily on high-cost renewable options could put modern energy services beyond the reach of many of their residents for years to come.

On the other hand, the energy infrastructure in some countries is so underdeveloped that it makes more sense to set up new renewable energy systems than to retool the existing infrastructure, or try to convert coal-based systems later on. Assuming catch-up growth and continuing rates of urbanization and industrialization, closing the gap between energy supply and energy demand in developing countries would require trillions of dollars of investment, even for low-cost options such as coal, and certainly well in excess of current energy investments in many developing countries. In the meantime, energy remains under-supplied and expensive in many areas, where people still rely primarily on traditional biomass fuels—wood, crop wastes and animal dung—for their energy needs.

For some rural areas, small-scale, decentralized renewable energy technologies now offer a cost-effective and sustainable approach to electrification. The cost per customer for connection to the national grid can be very high in remote or low-density rural areas, creating a natural advantage for local renewables. Still, any big push into low-emissions energy sources is likely to require massive investments in developing wind, hydro and other renewable energy sources and connecting isolated areas with the main national grid. Rising demand for liquid fuels and gases might potentially be met through the development of a modern biomass fuels industry, which could simultaneously increase farm and rural industry employment and income. Renewable energies could also generate backward linkages, as the search for low-carbon inputs would provide incentives to innovate and explore new activities. The existence of these alternatives for economic and social development underlines the need to include energy considerations in development planning.

The energy investment push

Figure II.4 depicts both the historical evolution of the energy system and one possible future development path towards decarbonization. This scenario

Figure II.4
Historical evolution of, and a possible future for, the global energy system, 1850-2100

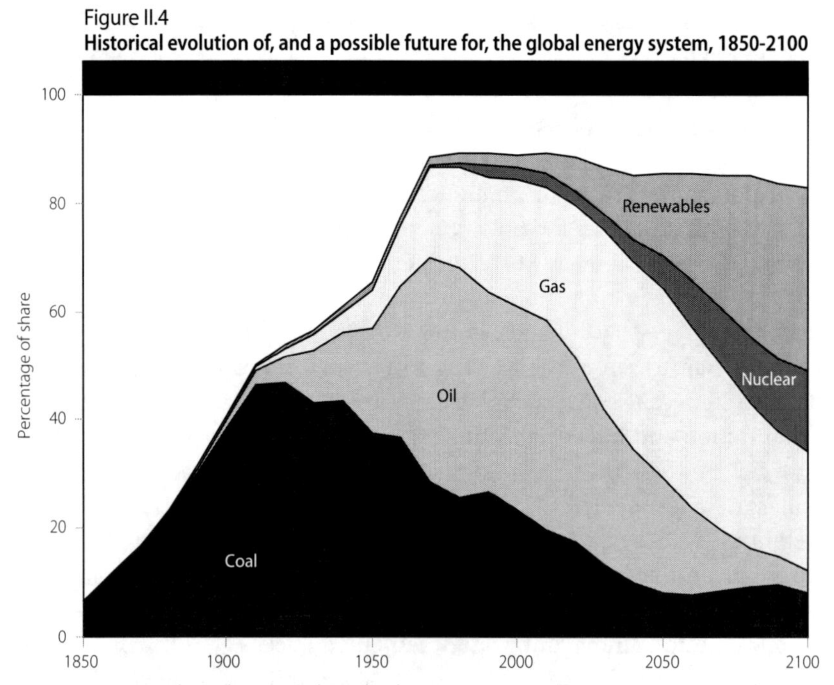

Sources: Riahi et al., 2007; Riahi and Nakicenovic, 2007; and International Institute for Applied Systems Analysis, 2009.

requires transformation of the global energy system, including new energy technologies and practices, as well as changes in lifestyles and behaviour. It describes a future world that stabilizes concentrations of greenhouse gases just above current levels and thereby limits global average temperature change to about 2° C by the end of the century. The climate change implied by such a scenario would be uneven across the world, and many regions might significantly exceed the 2° C global average. Hence, even a global temperature increase of 2° C can lead to considerable local vulnerabilities and disruptions of ecosystems, water supplies and coastal communities (see chap. III). Nevertheless, a 2° C world would be spared the most severe adverse (and perhaps also irreversible) consequences associated with higher rates of climate change. Under this scenario, the world would transition towards sustainability as well as economic convergence and the fulfilment of the Millennium Development Goals in most of the world. This is very much in line with the vision outlined in the previous chapter.

The uncertainties about the nature of technological change and its impact on the climate call for the adoption of innovations as early as possible, in order to ensure lower costs and faster, wider diffusion. As noted above, there appear to be significant global mitigation opportunities that would cost less than €60 per ton of CO_2e. This potential could be larger, especially if the price of fossil fuels increases. In 2008 and again in 2011, for example, oil prices were well over $100 per barrel. The IEA's *World Energy Outlook 2010* projects a rise in oil prices to $120 per barrel (in 2009 dollars) by 2025 and $135 by 2035 if no substantial energy policy changes occur, or $105 and $113, respectively, if countries pursue their pledged emissions reductions.[4] In light of recent experience, it is all too easy to imagine prices going much higher than that. Higher prices for oil and other fossil fuels increase the monetary value of the energy savings from mitigation measures, thereby decreasing the net cost of those measures.

However, as the 2008 oil price spike showed, a substantial increase in oil prices can create serious problems for energy-importing developing countries, with adverse impacts on fiscal solvency and increases in the costs of basic needs such as food, transportation and energy. A prolonged escalation in energy prices would be costly in developmental terms for many countries. Thus the adoption of a purely price-based carbon market strategy would require subsidies to developing countries to offset such adverse impacts. In addition, adequate domestic measures will be needed to ensure that subsidies reach poor and vulnerable groups.

Putting a price on carbon could also trigger some of the technological, institutional and behavioural changes required for effective emissions reduction, and given the low mitigation costs in developing countries, least-cost mitigation efforts would channel investment to these countries. However, these measures would have to be combined with a suite of compensatory policies to offset the social and economic costs of the price increase.

Technological learning and the change that it produces are essential for reducing mitigation costs and increasing mitigation potentials. To realize those benefits, "upfront" investments in new and advanced carbon-saving technologies will be needed, which would, after scale-up and adoption, lower mitigation costs and increase mitigation potentials. As chap. I suggests, these will initially have to be public investments. A recent study compared energy system investments throughout this century for two IPCC scenarios, A2 and B1.[5] The former is a typical "business-as-usual" scenario, with a rapid increase of greenhouse gas emissions leading to a likely global temperature change of about 4.5° C in this century; B1 corresponds to a slower emissions path that keeps this century's global temperature change under 3° C. The total investments under either scenario are in the range of $20 trillion through 2030 and are slightly higher for B1, owing to the initial build-up of capital-intensive energy systems. Meeting a 2° C target would imply still higher start-up investment. However, in the long term, beyond 2030, the capital costs of ensuring the more sustainable future are significantly lower owing to induced technological change and learning. Over the entire 21st century, energy system investments are about $40 trillion lower for the B1 scenario, compared to A2 (see figure II.5).

In other words, early upfront investments would enable potential reductions later as we move along the learning curves. Large investments would have to be made in developing countries in particular; indeed, again assuming that they will have the lowest costs, highest mitigation potentials, and largest opportunities for new markets, investments in the energy sector would likely be predominantly in those countries in the coming decades.

An integrated approach to the mitigation challenge

Energy security[6]

For many advanced countries, the future availability of oil has become a matter of concern and controversy. The IEA's *World Energy Outlook 2010* predicts a total oil supply of 107 million barrels per day by 2035 under its highest-consumption, highest-price scenario, and 99 million barrels if current emission reduction pledges are pursued. High oil prices will be

Figure II.5
Energy systems investment, 2000-2030

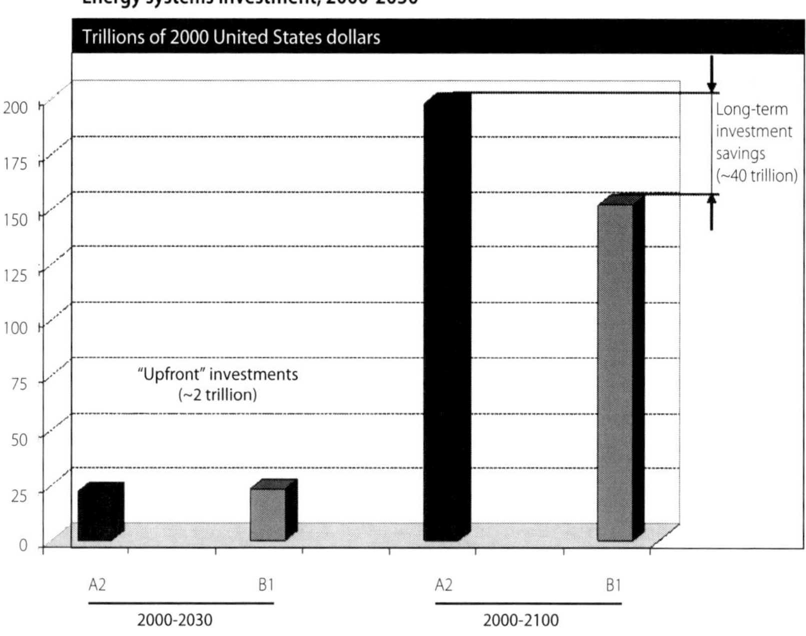

Sources: Riahi et al., 2007.
Note: Please see text for explanation of scenarios.

needed, it warns, to keep demand within the range of supply. The IEA also warns that the size of "ultimately recoverable resources" of conventional and unconventional oil "is a major source of uncertainty for the long-term outlook for world oil production".

Many energy experts hope that the supply of other basic fuels and energy sources—natural gas, coal, nuclear, hydropower—can be expanded even beyond current growth rates to compensate for possible shortfalls in the availability of oil. Still, without a radical shift in energy strategy, it will be difficult for these sources to fill the gap. This shift provides the opportunity to meet both climate and energy security goals in advanced countries.

Natural gas is the most attractive of the three fossil fuels from a climate perspective, because it emits the least amount of greenhouse gases per unit of energy. Nevertheless, gas is a finite commodity like petroleum, and many of the most prolific and easily accessible fields in North America, the North Sea and western Siberia have by now been largely depleted. Although new fields in eastern Siberia, offshore Iran, northern Alaska and Canada, and the Arctic Ocean await exploitation, the costs of developing these reservoirs will be substantially greater than the costs for those now in production, and it is not clear how many of them will attract the high levels of investment needed to bring them online. Additional supplies could be extracted from nonconventional sources in North America and elsewhere, but at the cost of significant and controversial local environmental impacts, which are not yet fully understood. In sum, while it is reasonable to expect some increase in the availability of natural gas in the years to come, it is unlikely to compensate fully for the eventual shortfall in petroleum supplies.

Coal is the most abundant of the fossil fuels. The technology for using coal to produce electricity is very well developed, and its relatively low cost has made it especially attractive to developing nations such as China and India as a source of electric power. The IEA's *World Energy Outlook 2010* projects that global coal use will rise by nearly 60 per cent between 2008 and 2035 under current policies (though only by 20 per cent if countries adopt policies to match their emission reduction pledges and stated plans). Conventional use of coal releases more CO_2 per unit of energy produced than oil or gas; an increase in coal use of this magnitude will result in a significant worldwide increase in CO_2 emissions, undermining global efforts to slow the rate of climate change. Hence efforts to stem CO_2 emissions would preclude a greater reliance on coal. Making coal viable in this context would require the development of cleaner coal-based technologies, in particular carbon capture and sequestration, which does not appear likely to be ready for large-scale deployment in the near future.

Another possible substitute for oil is nuclear energy. Because nuclear energy releases no CO_2 emissions, some see it as an attractive alternative to fossil fuels. Nuclear energy, however, also entails many risks, as the world was reminded by the Fukushima disaster in Japan—and previously, by the Chernobyl disaster. There are also severe, extraordinarily long-run problems of safe management and disposal of nuclear waste. All of this has kept costs high compared with other sources of energy, discouraging governments and private utilities from building many reactors. The tempo of reactor construction may pick up in the years ahead in response to rising demand for CO_2-free electrical power, but it is difficult to imagine a scenario entailing enough new plants to raise nuclear power's share of total world energy significantly above its current level of 6 per cent. The unfolding reactions to Fukushima may even reduce the long-term role of nuclear power below that level.

The only sustainable solution to energy insecurity and climate threats is the rapid development of climate-friendly renewable sources of energy, such as wind, solar, geothermal, and advanced biofuels. By the 2030s, renewable energy is projected to account for almost one-third of global electricity production, or one-sixth of global primary energy supply.[7] These projections could be revised upward in response to efforts by the European Union and the United States, but it will take a major investment push to lift the share of renewables by more than a few percentage points. After the sharp fall in oil prices in late 2008, many governments and utilities indicated that they would not be able to proceed with ambitious plans to develop new renewable energy projects because of inadequate funding.

To fully realize the potential of renewable sources of energy, a number of technological hurdles must be overcome. Before wind and solar power can be used more widely, for example, it will be necessary to find better ways to store energy and release it at night or when the weather is cloudy or windless. More efficient transmission systems are also needed to carry electricity from areas of greatest wind and solar energy production to areas of greatest demand. Likewise, new methods are needed to convert waste plant matter into ethanol, so as to spare food crops and other valuable species. Sources of energy such as geothermal, tidal power, hydrogen, and nuclear fusion will require a visionary approach and even greater scientific and technological advancement. These advances, in turn, will require investments that are not currently forthcoming—and some of today's "visionary" technologies may turn out to be unworkable at any price.

Until these multiple challenges of energy supply are addressed, the world will continue to experience persistent energy insecurity, which will make it difficult to overcome recurring economic insecurity. In that context, it is imperative to reduce the world's reliance on fossil fuels, and increase reliance on renewable sources of energy. For high-income countries this will require action on:

- ***Conservation:*** efforts to reduce the consumption of fossil fuels, especially oil. This means, among other things, driving less, driving more slowly, carpooling more, trading in gas guzzlers for fuel-efficient cars, improving the energy efficiency of homes, businesses and electrical appliances of all types, and expanding public transportation.
- ***Innovation:*** developing ever more fuel-efficient vehicles, factories, appliances, and heating systems; moving to gas/electric hybrids, plug-in hybrids and all-electric cars; improving the efficiency and utility of wind and solar power; developing advanced biofuels derived from non-edible plants.
- ***Investment:*** greatly increasing public and private investment in energy efficiency,[8] renewable energy sources and public transportation; creating

financial incentives for the development and utilization of energy alternatives, such as green bonds and a cap-and-trade system for carbon emissions.

Efforts along all fronts must start immediately if real progress is to be made. Renewable-energy investments have already risen significantly, from $33 billion in 2004 to a record $211 billion in 2010. China is a major driver of this growth, investing nearly $50 billion in 2010 alone. Even the mistakes of European policy—unsustainably generous solar power subsidies, later rescinded—have spurred growth of capacity and reduced prices in photovoltaics, for which small-scale installations are now increasing rapidly.[9] But a real transformation of energy systems will require far more; the sustainable scenario depicted in figure II.5 would require at least $1 trillion per year over the coming decades.

Energy access

Given the overall low level of energy consumption in developing countries, the concept of energy security is predictably somewhat different than in more advanced economies. Access to modern energy services is unequal, notably between rich and poor and between urban and rural areas. Indeed, about 2 billion people, almost one third of the world population, have no access to modern energy; about 1.6 billion have no electricity, while 2.4 billion cook with traditional forms of biomass. As the United Nations Development Programme has noted, limited energy access is an important contributor to high levels of poverty in some sub-Saharan African countries. Yet this limited-access population makes up a vast potential future market for energy. Most live in rural areas; about 260 million are estimated to be urban dwellers. Expanding energy access to these populations would have a positive effect on economic activity and development.

Assuming an average connection cost for those excluded at $1,000 per household (and a 20-year phase-in period) yields global investment needs of some $25 billion per year over the next 20 years. This is a huge sum for the poorest of the developing countries, but a humble one compared with other requirements for global climate and development solutions. It pales beside the hundreds of billions recently spent by many industrialized countries to rescue the financial sector, automotive industry and other sectors of the economy. Still, only about $4 billion, or 4 per cent of the total Official Development Assistance provided each year, now goes to energy needs.

Capacity expansion

Scenarios of future energy development also assume a substantial improvement of energy services, with developing countries as the largest future energy markets. Capacity expansion in the South (developing countries) is expected to be double that in the North (industrialized countries) over the coming decades, demonstrating how significant growing energy markets will be in the developing world. Capacity replacement is much larger in the North because of its huge existing stock of power plants and their substantial ageing. In business-as-usual scenarios with continuing reliance on fossil and nuclear energy, especially coal in the United States, China, India and Russia among others, the total new capacity to be installed is almost 50 terawatts electric (TWe) or 12 times the current global installed capacity. (A terawatt is a million megawatts, or a trillion watts; a terawatt of generating capacity is, roughly speaking, 1,000 large coal or nuclear plants, or 400,000 wind turbines). Even under these scenarios, developing countries would add renewable capacity through 2030 equivalent to all the power plants in the world today, and half as much again in additional nuclear plants. This presents important investment opportunities for the private sector, but in terms of climate mitigation, it would be dwarfed by the expansion of traditional fuel sources.

Climate stabilization scenarios reduce, but do not entirely eliminate, the need for capacity expansion and replacement in coming decades; the exact amount that is needed depends on the extent of energy efficiency initiatives, among other factors. Moderate stabilization targets, assuming substantial success in energy efficiency and conservation programmes, still entail many terawatts of new and replacement capacity—with the bulk of the investment in the South.

With so much expansion of energy systems slated to occur, there is a potential risk of locking in traditional technologies if the new investments in generating capacity do not use the best technologies. This risk should provide a huge incentive to attract capital to renewable energy and to support its extension to the developing world. On the other hand, there are real possibilities in developing countries of leapfrogging to the most advanced technologies. There is obvious potential for a virtuous growth circle (which also meets the climate challenge), in which a big public investment push in mitigation action leads to crowding in of private investment, technological upgrading and productivity growth.

Feed-in tariffs and renewable portfolio standards

Among the policy options that might promote a rapid transition to renewables, there has been particular interest in the feed-in tariff—a policy that obligates utility companies to purchase, at a legally mandated price per kWh (or "tariff"), energy "fed into" the grid from renewable sources by any individual or organization.

Feed-in tariffs have been used for over two decades and are now on the books in at least 45 countries and U.S. states. They were first developed in the United States under the Public Utility Regulatory Policies Act of 1978, which allowed connection of renewable generators to the grid and specified that they should be paid for the cost of generation that they avoided. In response, different states developed contractual arrangements, called "standard offer contracts", to offer to renewable generators. The first was California's Standard Offer No. 4, in 1984, which fixed the amount to be paid per kilowatt-hour for a long period (generally 10 years, in the context of a 30-year contract). The rate was based on the long-term avoided cost of conventional generation. These contracts led to the establishment of 1,200 megawatts of new wind power plants by the late 1980s, which have contributed about 1 per cent of California's consumption for more than two decades. While the legal forms have changed, California has continued to experiment with other feed-in tariff options.[10]

Germany implemented its Stromeinspeisungsgesetz (the "law on feeding in electricity") in 1991. Germany based its tariffs upon a fraction of the retail rate (that is to say, the price at which electricity was sold to consumers), not the wholesale rate (the cost at which utilities purchased electricity from other generators). Wind energy and solar energy were paid 90 per cent of the retail rate, and hydroelectric plants were paid 80 per cent of the retail rate.

However, even these rates were not sufficient to attract adequate financing. This was corrected in Germany in 2000 by the stipulation that renewable sources of electricity would have priority access to the grid for a host of environmental, social and economic reasons. Germany also set different tariffs for different technologies (based on the respective cost of generation plus a reasonable profit) and guaranteed them for 20 years. Many other countries have followed this model, called Advanced Renewable Tariffs.

For residential rooftop solar photovoltaic (PV), for example, Germany's 2004 law offers €0.57/kWh, much more than for other sources. Similarly, the Canadian province of Ontario recently revised its laws to offer standard contracts differentiated by technology, size and application, including C$0.80/kWh for residential rooftop solar PV.

In developing countries, the costs of most renewable options are far higher than the average retail price of electricity, which is often held quite low because so many people cannot afford higher rates. This creates a disincentive for producers, who fear future policy changes in case of large-scale uptake of renewable energy generation. In this regard, a feed-in tariff option can be successful in developing countries only if it is backed by an international guarantee and internationally funded subsidies for low-income consumers.

The feed-in tariff is just one of several policy options for inducing investments in renewable energy, which may be used alone or in combination. Other options include renewable portfolio standards, which require utility companies to supply a specified share of electricity from renewable sources; price mechanisms, which raise the price of carbon-based energy, for example, through a carbon tax or a cap-and-trade system; and direct or indirect support for the renewable sector through allocation of funds for research and development, subsidized credit or land, or even direct public investments.

In the United States, California—long an environmental leader, but also, by virtue of its size, one of the world's largest greenhouse gas emitters—has adopted several policies that have driven change in the private sector. Under the California Renewables Portfolio Standard (RPS), for example, requiring utilities to procure 20 per cent of their power from renewable sources by 2010, renewable capacity grew by 1,702 MW from 2002 to 2010, according to the California Public Utilities Commission. Utilities still fell short of the standard, procuring only 17.9 per cent of their power from renewables (excluding large hydroelectric) in 2010, according to state data, but that is still almost five times the share of electricity generated from renewables in the United States as a whole in 2009 (3.6 per cent). Another 889 MW of renewable capacity was expected to be added in California in 2011, and the state has approved a new RPS with a goal of 33 per cent renewables by 2020.[11]

Research and development

Research, development and deployment (RD&D) are vital to the improvement of performance and the lowering of costs in the early stages of any technology. RD&D expenditures are ultimately a small factor in the cost improvements of technologies that have already found commercial niche markets, but in the earlier stages, RD&D accounts for a large share of performance improvements and cost reductions.

Research and development efforts that lead to diffusion of new and advanced technologies can address the double challenge of providing access to modern technology for those who have been excluded, while allowing further development opportunities among the more affluent. In the energy area, this could imply a shift from traditional fuels, in the case of those who are excluded from access, to clean fossils and modern renewable energy; and, in the more developed parts of the world, a shift from fossil energy sources to carbon-free and carbon-neutral energy services. In all cases, this means a vigorous improvement of energy efficiencies, from supply to end use, expanding shares of renewables, more natural gas and less coal, vigorous deployment of carbon capture and storage, and—depending on social acceptability, economic viability, and resolution of environmental risks—perhaps also nuclear energy.

All of these transformational changes in the energy system need to be empowered by vigorous RD&D efforts, investments, removal of barriers, provision of information and capacity-building. The latest trends are not entirely positive. In 2010, public sector support for R&D on renewables rose by 121 per cent, to $5.3 billion, but corporate R&D dropped by 12 per cent, to $3.3 billion. There are risks that progress on renewables could be hindered by fear of government cutbacks, plummeting natural-gas prices, and scepticism about the future of the market.[12] Yet, much greater investments are needed to assure the timely replacement of energy technologies and infrastructures.[13]

Conclusion

A more sustainable future requires large upfront investments, likely more than $1 trillion per year from now to 2030, primarily targeting developing countries. Achieving a transition towards more sustainable development paths will also require substantial and complementary investment in energy RD&D.

Not only would these investments lead to carbon-leaner energy systems and a more sustainable development path, but in the long run (to 2050 and beyond), they would reduce costs compared with business-as-usual alternatives. The reason is that the cumulative nature of technological change translates the early investment in a lower-carbon future into lower costs of the energy systems in the long run, along with the co-benefits of stabilization. This points to the need for radical change in energy policies in order to assure that the investment effort will be adequate to our common future, and to promote accelerated technological change in energy systems and end uses.

Notes

1 World Resources Institute, 2010. 'Climate Analysis Indicators Tool.' CAIT 8.0. http://cait.wri.org/.
2 Hirschman, Albert O., 1958. *The Strategy of Economic Development*. Yale University Press, New Haven, CT.
3 International Energy Agency, 2010. *World Energy Outlook 2010*. Paris. http://www.worldenergyoutlook.org/2010.asp.
4 Ibid.
5 Riahi, K., Grübler, A. and Nakicenovic, N., 2007. 'Scenarios of long-term socio-economic and environmental development under climate stabilization.' *Technological Forecasting and Social Change*, 74(7). 887–935. doi:16/j.techfore.2006.05.026.
6 This section draws substantially from Klare, M., 2008. 'Persistent energy insecurity and the global economic crisis.' Paper presented at the panel discussion on Overcoming Economic Insecurity, Second Committee, United Nations General Assembly, 11 November.
7 Electricity projection is from IEA (ibid.), New Policies Scenario; primary energy supply is based on an IPCC review of 164 scenarios from 16 large-scale integrated models that found more than half projected that renewables (including biomass) would contribute 17 per cent or more of the global primary energy supply in 2030: Intergovernmental Panel on Climate Change, 2011. *IPCC Special Report on Renewable Energy Sources and Climate Change Mitigation*. O. Edenhofer, R. Pichs-Madruga, Y. Sokona, K. (coordinating lead authors); Seyboth, P. Matschoss, S. Kadner, T. Zwickel, P. Eickemeier, G. Hansen, S. Schlömer, and C. von Stechow (eds.). Cambridge University Press, Cambridge, U.K., and New York.
8 As argued in a recent United Nations report, much if not most of the reduction of greenhouse gas emissions to meet climate change mitigation targets can be achieved by gains in energy efficiency. This would require giving sufficient priority to development of energy-efficient end-use technologies (automobiles, electrical equipment, heating systems, etc.). Hence, a mere focus on developing clean and renewable energy source will not suffice. See United Nations, 2011. *World Economic and Social Survey 2011: The Great Green Technological Transformation*, New York. http://www.un.org/en/development/desa/policy/wess/wess_current/2011wess.pdf
9 United Nations Environment Programme and Bloomberg New Energy Finance, 2011. *Global Trends in Renewable Energy Investment 2011: Analysis of Trends and Issues in the Financing of Renewable Energy*. UNEP Division of Technology, Industry and Economic, Frankfurt School-UNEP Collaborating Centre for Climate & Sustainable Energy Finance, and Bloomberg New Energy Finance. http://fs-unep-centre.org/publications/global-trends-renewable-energy-investment-2011.
10 Rickerson, W., Bennhold, F. and Bradbury, J., 2008. *Feed-in Tariffs and Renewable Energy in the USA—a Policy Update*. Report sponsored by the Heinrich Böll Foundation, the North Carolina Solar Center, and the World Future Council. http://archives.eesi.org/files/Feed-in%20Tariffs%20and%20Renewable%20Energy%20in%20the%20USA%20-%20a%20Policy%20Update.pdf
11 California Public Utilities Commission, 2011. *Renewables Portfolio Standard Quarterly Report—1st Quarter 2011*. Sacramento, CA. http://www.cpuc.ca.gov/NR/rdonlyres/62B4B596-1CE1-47C9-AB53-2DEF1BF52770/0/Q12011RPSReporttotheLegislatureFINAL.pdf

Data for United States as a whole are from Gelman, R. and Kubik, M. (ed.), 2010. *2009 Renewable Energy Data Book*. National Renewable Energy Laboratory, U.S. Department of Energy, Washington, D.C. http://www1.eere.energy.gov/maps_data/pdfs/eere_databook.pdf

12 United Nations Environment Programme and Bloomberg New Energy Finance, 2011 (ibid.).

13 See for further discussion and comparison of estimates, United Nations, 2011. *World Economic and Social Survey 2011*, ibid.

Chapter III
The adaptation challenge

Introduction

The previous chapters have argued that living standards can be raised in developing countries without jeopardizing climate change mitigation efforts. It is clear, however, that the development path followed by today's rich industrialized countries can no longer serve as a model for catch-up growth. Rather, powering industrial expansion, rapid urbanization and population growth in the developing world will require a big push into cleaner and more efficient technologies, above all in the production and consumption of energy.

But even if the world shifts quickly onto a low-emissions path, the accumulation of greenhouse gases in the atmosphere from past emissions means that a noticeable increase in global temperatures is unavoidable and will bring serious environmental damage, including increased droughts, rising sea levels, ice-sheet and snow-cover melting, and extreme weather events. These phenomena will, in the coming decades, threaten and destroy livelihoods around the globe, especially for already vulnerable populations. As our knowledge of climate change deepens, scientists are becoming increasingly alarmed about the potential scale of environmental damage from what were previously considered manageable changes in global temperatures. The threats to livelihoods and security are also likely to be that much greater.

For many developing countries, environmental constraints and shocks are already part of a vicious development cycle that traps them in poverty, undermines their resource base and restricts their capacity to build resilience to future shocks. The problem is sure to become even more challenging with global warming. Poor health-care systems, lack of infrastructure, weakly diversified economies, missing institutions and soft governance structures expose poorer countries and communities not just to potentially catastrophic large-scale disasters but also to a more permanent state of economic stress from higher average temperatures, reduced water resources, more frequent flooding and intensified windstorms. These stresses will likely increase the risks of food and income insecurity and further expose the inadequacy of health care, sanitation, shelter and social infrastructures.

Adapting to climate change must be a central component of any comprehensive and inclusive climate agenda. Several international funds have been set up to finance adaptation measures in developing countries, but they are woefully inadequate. Scaling up these funds is the first challenge in the adaptation agenda. A second challenge is that adaptation is still seen primarily as an environmental issue, and there is a tendency to compartmentalize climate change policies and isolate them in environmental ministries.

Even when adaptation measures have been linked to a development strategy, the focus has often been either on poverty alleviation (with the goal being to promote stronger safety nets and innovative insurance mechanisms for vulnerable groups and sectors) or on business opportunities (from strengthening climate-related markets). These actions have a role in a more integrated strategy, but they cannot frame it. Rather, this chapter argues that increased investment, improved access to finance, and strengthened regulations and institutional capacity are, as with climate mitigation, at the heart of confronting the adaptation challenge in most developing countries. Indeed, synergies between adaptation and mitigation strategies need to be explored much more fully, as an integral part of low-emissions, high-growth development pathways in countries vulnerable to climate change and shocks.

Adaptation and vulnerability

Mitigation is directed at slowing the growth of future emissions of greenhouse gases and eventually reducing their stock to a level consistent with manageable and stable temperatures. Adaptation is about mobilizing resources and devising policy strategies for coping with the unavoidable negative impacts of higher temperatures. This is not a challenge that is altogether new. Throughout history, human societies have shown an extraordinary capacity for adapting to climatic changes. However, the threats posed to security and livelihoods by anthropogenic global warming, like the appropriate responses, are likely to be unprecedented.

Climate change and vulnerability

Despite considerable variation in scientific estimates of the possible impacts of climate change, concern continues to grow over catastrophic risks to the planet's ecology and to life in general. For example, James Hansen and others argue that the eventual temperature increase from a doubling of atmospheric carbon dioxide (CO_2) is more likely to be 6° C, rather than the 3° C assumed by both the Intergovernmental Panel on Climate Change (IPCC) and the *Stern Review*.[1] Many scientists estimate that global warming of 4° C or more is increasingly likely in this century; one recent analysis finds that under the IPCC's A1FI scenario (a high-emissions business-as-usual pathway), 4° C above pre-industrial temperatures would likely be reached by the 2070s or sooner; under the A2 scenario, that temperature threshold would likely be reached by the 2090s.[2]

The damage resulting from climate change will not be felt uniformly across countries and communities (see chap. I). The United Nations Development Programme notes that climate models show large losses in productivity for food staples in sub-Saharan Africa and South and East Asia, for example, and projects that 600 million people could face acute malnutrition by the 2080s due to climate change.[3] By contrast, there will be some areas of the world that may experience small benefits, for instance, with regard to mortality rates and crop yields, provided that global temperature increases do not greatly exceed 2° C. (It should be noted, however, that the latest agricultural research suggests smaller gains and more widespread losses from even the earliest stages of warming.) Should temperatures rise more than about 2° C, the resulting proliferation of threats could rapidly heighten existing vulnerabilities.

New data on the melting of mountain glaciers and the ice sheets of the Arctic and Antarctic point to an increased likelihood of a significant rise in sea levels, which could lead to serious threats to several big cities such as New York, London, Tokyo, Dhaka, Shanghai, Mumbai and Rio de Janeiro. Similarly, in the Andean cordillera, melting of glaciers threatens the water supply and livelihoods of at least 30 million people (see box III.1). Worldwide, the livelihoods of about 500 million people depend on glacier-fed rivers. Sea-level rise and storm surges are also a direct threat to hundreds of millions of people in low-lying coastal zones.

Communities' adaptive capacity, of course, is also a major factor. Developed countries can draw on financial resources and institutional strengths to bounce back from environmental shocks and bolster their resilience to future impacts. This is not the case in most developing countries. In Africa, for example, where drought and famine are already serious problems, a UNFCCC report notes, communities' ability to respond to climate change will be hindered by poverty, illiteracy and lack

Box III.1: The multiple threats to livelihoods from climate change: the Andean case

The impacts of climate change are cumulative and are closely linked to other vulnerabilities, often in a dangerously reinforcing manner. This is clearly illustrated by the accelerated melting of mountain glaciers, which are a critical source of livelihoods for about 500 million people worldwide and essential contributors to regional and global biodiversity.

Most of the world's tropical glaciers are located in the Andean mountains of Peru, Bolivia and Ecuador, where melting threatens the water supply and livelihoods of at least 30 million people. Over one fifth of the surface of 18 mountain glaciers in Peru has already melted over the past 35 years, while most of the lower-altitude Andean glaciers are expected to diminish substantially during the next 10-20 years.

Direct impacts of this trend are being felt in large cities in the region, which depend on glacial run-offs for their water supply. Quito draws 50 per cent of its water supply from the glacial basin, and La Paz, 30 per cent. The loss in volume of the glacier surface of Peru, equivalent to 7,000 million cubic metres of water (about 10 years of water supply for Lima), has meant a reduction by 12 per cent of the water flow to the country's coastal region, which is home to 60 per cent of the population of Peru.

As glaciers retreat, the capacity to regulate water supply through run-off during dry and warmer periods and to store water in the form of ice during wet and colder periods is being lost, which in turn puts agriculture and power generation at risk. Without sufficient run-off, pasture land upon which to raise livestock and continue small farming will be insufficient. As the cultivation of local staples such as potatoes and quinoa is likely to dwindle, farmers may have to resort to planting costlier crops that need chemical fertilizers.

Moreover, most Andean countries are also dependent on the glaciers for hydroelectric power generation, which accounts for 50 per cent of the energy supply in Bolivia and 70 per cent or more in Colombia, Ecuador and Peru. With rising temperatures, energy generation will be diminished in areas where water basins depend on glaciers, increasing the need to invest in additional power capacity and, as in Peru, explore thermal-based power options.

Source: The World Bank, 2008.

of skills, weak institutions, limited infrastructure, lack of technology and information, poor access to resources, low management capabilities, armed conflicts, and other factors.[4] The problems are even worse, meanwhile, for small, low-lying island nations such as the Maldives and Kiribati, where the climate threat is already so daunting that relocating their populations may be the only viable response.[5]

Climate change will also compound the interrelated threats faced by vulnerable communities. For instance, the number of outbreaks of tropical diseases is likely to be larger in areas with more frequent and severe heat waves, while the incidence of water-related diseases is likely to rise in areas with more floods. Increased hurricane activity will also lead to an increase in respiratory diseases (for example, influenza), in particular when emergency shelter is inadequate and in areas with little or no medical assistance. As Harvard Medical School researcher Paul Epstein warns, climate change "could devastate public health by midcentury".[6] In addition, people will lose their livelihoods, lack adequate shelter and face food insecurity—with the worst impacts on vulnerable groups such as children, older persons and women.

The 2008–2009 winter drought in northern China, the worst in 30 years, provides an example of the compounding effects of multiple climatic shocks. Due to lack of rain and snow since November 2008, China's Ministry of Water Resources reported in early February 2009, about 3.7 million people and 1.9 million large animals had limited access to drinking water in northern China. The drought damaged about 180,000 hectares of farmland, affecting about 40 per cent of China's winter wheat crop and decreasing the total wheat harvest by about 5 per cent. Officials declared a state of emergency and allocated about 400 million yuan for drought relief.[7]

Adaptation and development

Reduced vulnerability to natural hazards is strongly correlated with income levels, and reflects changes in economic and social structures as countries diversify away from reliance on agricultural activities, establish stronger institutional networks, and begin to build more effective welfare states. Adaptation to actual or expected climate change will inevitably involve large investments to protect existing activities and livelihoods, facilitate needed adjustments, and exploit potential opportunities. Adaptation can emerge spontaneously as individuals and communities respond to repeated shocks or incremental changes in their surrounding environment. However—particularly when the changes are on a larger scale—lasting solutions will require deliberate policy decisions and public action, based on scientific research, assessment of previous crisis episodes, and consultations with local residents and grassroots groups.

Adaptation may be particularly hard for populations that are already vulnerable due to low levels of economic and human development. Experience with disasters shows how difficult it is to break out of poverty and insecurity that have been exacerbated by climate shocks, and how difficult it is to judge how much of the resulting impact can be attributed to "normal" economic versus "abnormal" climate factors. This also underscores the interrelated nature of climatic and development-related pressures in adaptation. Not only can the damages be much larger than the resources available to provide protection, but poor countries may have difficulties mobilizing those resources and investing in effective adaptation measures.

Even in developing countries that have managed to foster sustained growth, vulnerability to shocks, both internal and external, remains a persistent concern for policymakers. Poor neighbourhoods, even in developed countries, are more at risk from shocks, including climatic shocks, because they have fewer coping resources and are inadequately served by basic services that are taken for granted in more affluent areas. Economic growth alone does not guarantee that needed adaptation measures will be taken; the United States had ample warning that New Orleans was at risk from a major hurricane, but did not construct adequate levees before Hurricane Katrina devastated the city in 2005.

In responding to the adaptation challenge, policymakers can usefully draw on the experiences of developing countries in adjusting to exogenous economic shocks. Perhaps the strongest conclusion that emerges from those experiences is that local circumstances and capacities have a profound influence on outcomes, and that policy responses should be tailored accordingly. However, some general lessons can also be drawn, such as:

- If countries are left to adapt on their own, they will likely be forced to squeeze down incomes, which would result in a prolonged and destabilizing adjustment process, increasing poverty levels, damaging long-term growth prospects and adding to further vulnerabilities.
- Economies that are more diversified (both structurally and spatially) tend to show greater resilience to external shocks and recover more quickly, as do economies that are strongly integrated both internally and externally.
- Societies with less inequality are better able to manage shocks by equitably distributing the burden of adjustment and avoiding the possibly dangerous conflicts that adjustment can trigger.

Adaptation is very much a *local* challenge which will require locally appropriate strategies and mechanisms. However, in general, economic development is the most reliable insurance against climate impacts. On the whole, populations that have access to adequate food, clean water, health care and education are better prepared to deal with a variety of shocks, including those from climate change. Access to adequate resources to invest in adaptive capacity, including human and social capital, determines how resilient countries and communities are likely to be. In addition, access to technologies and know-how will play an important role in strengthening adaptive capacity. Thus the ability of decision makers to mobilize and manage resources and to engage in difficult trade-offs involving their use will be an essential component of the response to the adaptation challenge.

Many developing countries remain heavily dependent on natural resource-based activities and are likely to be seriously threatened by projected climate changes. Communities and countries that primarily produce and export low value-added agricultural goods and primary commodities are typically found at the lower end of the development ladder and are vulnerable due to small market size, heavy import dependence, low technological capacity and other factors. Food security remains a basic challenge, particularly where agriculture is dominated by smallholder production, productivity is low, and support services are poorly developed. Climate change is certain to exacerbate these problems.

Other developing countries are transitioning to more urban and economically diversified economies and must cope with new risks and interrelated shocks. By 2030, it is estimated that 60 per cent of the world's population will reside in urban areas, compared with 47 per cent in 2000, according to the United Nations Human Settlements Programme (UN-Habitat).[8] Cities are also national and regional economic hubs and play crucial roles in the globalization process. The policy challenges accompanying this transition can be compounded by increasing insecurity and inequality, as new urban residents may find themselves forgoing the minimal levels of protection offered in rural communities, while receiving little or no government support.

Overall, in the absence of more effective adaptation strategies, rising global temperatures are likely to exacerbate the differences in vulnerability between rich and poor—both internationally, and within countries. This is a concern for the international community both in its own right and because in an increasingly divided and unequal world agreement on an international framework for tackling climate change will be even more difficult to achieve.

The limits of existing policy frameworks

The economic stabilization and structural adjustment programmes implemented in many developing countries over the past three decades have done little to reduce vulnerability. Adopted in response to a series of large shocks in the late 1970s and early 1980s and to the ensuing debt crisis, those programmes aimed to remove structural and institutional impediments to growth and create more stable and resilient economies. They typically gave a much greater role to market forces and reduced that of the state, including capacities for providing public services. One prominent aspect of this shifting emphasis was fiscal retrenchment and the accompanying decline in public investment across much of the developing world. As a consequence, even with greater macroeconomic stability, private investment was insufficiently supported through infrastructure and basic services, thereby limiting productivity growth and economic diversification. In many instances, income-earning capacities were not improved and sometimes even fell, through premature de-industrialization, wage compression and the informalization of economic activity.[9]

Towards the end of the 1990s, a second generation of adjustment programmes added good governance and poverty reduction to the reform agenda, in part to deal with perceived policy slippages but also in response to the adverse impact of the earlier measures. These efforts have placed a greater emphasis on participation and ownership in the design of programmes, culminating in the preparation of Poverty Reduction Strategy Papers (PRSPs) which have become the main policy vehicle for allocating bilateral grants and concessionary loans and for advancing debt relief. However, the PRSPs have mostly left intact the economic reforms of the first-generation programmes, have done little to seriously assess the impact of major macroeconomic and structural measures on the poor, and have failed to establish a more integrated approach to economic and social challenges. In particular, as a 2002 United Nations Conference on Trade and Development analysis found, they have continued to promote unduly restrictive macroeconomic policies to the detriment of investment-led growth and diversification strategies, denied the contribution of industrial and technology policies towards supporting such strategies, and adopted a one-size-fits-all approach to integration into the international economy.[10]

In order to adapt to a warming world, developing countries will need new policies that build robust links among investments, growth and diversification, allowing them to make progressive adjustments to climatic changes and to strengthen national resilience to climatic shocks.

The impacts of climate change

As noted, the damage from climate change will not be felt uniformly. Some damages will be gradual, such as those caused by sea-level rise or the spread of drought. Others will be inflicted suddenly, in the form of more frequent and intense natural disasters. Some threats will be confined to specific sectors, while others will have a systemic impact. Moreover, while the impacts will have ramifications across all countries and regions, their intensity will often be quite localized, with some communities and countries much more exposed than others.[11] In general, most of the imminent adverse impacts on livelihoods are expected to be felt in developing regions, where drought (Africa) and flooding (parts of Asia) are already serious threats. On the other hand, heat spells might challenge water security in some developed regions, for instance, in Australia and New Zealand, particularly if temperatures rise by more than 2° C.

Agriculture and forestry

Globally, more than one third of households rely on agriculture for their livelihoods; in sub-Saharan Africa, the proportion is over 60 per cent. Moreover, in many poorer countries, primary products are a major source of foreign-exchange earnings and provide important inputs into fledgling manufacturing activities. While the economic weight of this sector is expected to decline further over the coming decades, improving agricultural performance is essential for food security and for sustained economic growth, particularly at lower levels of development.

The net impact of climate change on global agricultural production remains uncertain. There are regional variations in the expected pace of warming, but the agriculture and forestry sectors in developing countries of all regions are particularly vulnerable to climatic shifts, as even small changes in temperatures and precipitation levels can disrupt growth cycles and yields.

Significant reductions in the average yield of key staples and protein sources, and increased flood risks and consequent damage to assets are a few of the most adverse effects of climate change on developing regions. In contrast, warming and a general increase in rainfall are likely to lead to increases in crop productivity in northern and central Europe, particularly as some crops that are traditionally grown in southern Europe will become viable farther north. However, newer research suggests that the benefits in temperate and cooler regions may not be as great as initially estimated, if there are benefits at all. The economist William Cline, for example, has projected that by the 2080s, a business-as-usual climate scenario would reduce world agricultural output by 16 per cent without any estimate for carbon fertilization, or by 3 per cent with carbon fertilization effects included. For the United States, Cline projects an agricultural output loss of 6 per cent without carbon fertilization, or a gain of 8 per cent with it.[12] Other recent research implies even greater yield losses for the United States, driven by the rapid increase in the number of extremely hot days during the growing season.[13]

For developing countries, not only will the impacts be more uniformly negative, but their greater reliance on agriculture and the vulnerabilities of small-scale producers will make it even harder to cope. In many developing regions, growing seasons will shorten, areas suitable for agriculture will decline, and land degradation will intensify—especially along the margins of semi-arid and arid areas. Moreover, heat-related plant stresses will contribute to reduced yields in staple crops. In sub-Saharan Africa, one recent study projected, yields will drop by 22 per cent for maize, 17 per cent each for sorghum and millet, 18 per cent for groundnut, and 8 per cent for cassava by mid-century.[14] Likewise, extreme wind, storms, and turbulence are expected to decrease fish catches in several countries; the effects of ocean warming and acidification will only make this problem worse, reducing the productivity of fisheries almost everywhere.

Food security and rural livelihoods are closely linked to water availability and use. Scarcity of freshwater already threatens livelihoods linked to agriculture and forestry in an estimated 40 per cent of rural areas worldwide, and the heightened threat from climate change introduces the risk of far greater damage, increasing the likelihood of social conflict and triggering large-scale migration. Rising sea levels may also lead to salinisation of rivers, which further increases freshwater stress.

Moreover, many developing countries are lacking irrigation and modern agricultural inputs such as fertilizers, herbicides and insecticides. Under these conditions, yields are low, and climate change could have disastrous consequences for food security. In Mali, for instance, the proportion of the population at risk for hunger could increase from 34 to over 70 per cent by the 2050s, one study estimated.[15]

Forests cover about 30 per cent of the global land surface and contribute to the livelihoods of more than 1.6 billion people, providing food, fuel for cooking and heating, medicine, shelter and clothing, according to the Food and Agriculture Organization of the United Nations (FAO). An estimated 60 million highly forest-dependent people live in the rainforests of Latin America, South-East Asia and West Africa; another 350 million people directly depend on forest resources for subsistence or income, and 1.2 billion people in developing countries rely on trees on farms. In much of rural sub-Saharan Africa, forest foods are a regular part of the diet, and some farmers earn more than half their cash incomes from selling forest products such as wild honey, charcoal, fuelwood and wild fruits.[16]

Rising temperatures, shifting precipitation patterns and increasing emissions are likely to have significant impacts on forest growth, though the extent to which they will be positive or negative is not yet entirely clear. Indirect impacts such as the intensity of forest wildfires, invasions of insects and pathogens, and extreme weather events will definitely do harm. Overall, climate change is expected to both increase global timber production and shift supply locations from temperate to tropical zones and from the northern to the southern hemisphere. While this will lead to an increase in the trade in forest products, the benefits are likely to be unevenly distributed and are expected to adversely affect people in extreme poverty who depend on forests for their livelihoods.

Urban environments

Cities, meanwhile, are already home to more than half the world's people, and as noted earlier, the urban share of the population is growing rapidly, especially in developing countries. Urbanization is a major driver of climate change, and climate change will also have a significant impact on urban environments, adding a dangerous feedback loop to growing urban stresses.

Much of the urbanization in developing countries is unplanned and poses massive challenges, even without taking heightened climatic threats into account. These include health problems linked to air pollution and high population density, problems associated with transportation and inadequate infrastructure, personal safety problems linked to high levels of criminal activity, and deficient access to and provision of social services. Climate change is likely to exacerbate all these problems. Increased migration from rural areas affected by climate change will also put pressure on cities, overloading their already strained infrastructure and services.

While there are many climate-related hazards facing urban areas, coping with an increased incidence of natural hazards may pose the most immediate challenge. For instance, unplanned urban settlements, especially slums, often materialize in high-risk areas, such as river banks and unstable hill slopes. While slum dwellers may be able to cope with occasional shocks, more frequent flooding of greater magnitude would likely force them to resettle, possibly pushing them into greater poverty and increased climate exposure.

In the absence of any forward-looking strategy, an estimated 1 billion people are already at risk from hydro-meteorological hazards, and that figure is predicted to increase to 1.4 billion people by 2020, according to UN-Habitat.[17] More frequent and more intense rainfall will, for instance, increase the risk from landslides and the threat from water inundation. In fact, poor drainage is already a serious issue in many cities, particularly in developing countries, and climate change increases the risk of both flooding and disease.

Health and water security

The need to adapt to difficult environmental conditions has been a perennial challenge for human societies. Now, with warming on an accelerating trend, the impacts of climate change on health and water security merit particular attention.

The range of health risks from climate change is likely to be considerable, with all parts of the globe affected, as the unprecedented number of deaths in Europe from recent heat waves has demonstrated. However, health vulnerability is closely linked to other vulnerabilities. The people most vulnerable to climate change are those who already lack access to adequate health services, and the greatest factor determining the intensity of impacts across regions is not the severity of climate change, but variation in the magnitude of pre-existing health problems.

Many of the most important infectious diseases are highly sensitive to both temperature and precipitation conditions, so global warming will increase the prevalence of vector-borne diseases, particularly in poor countries. The World Health Organization notes that many major killers, such as diarrhoea, malnutrition, malaria and dengue are "highly climate-sensitive" and expected to worsen over time; already, it estimates, the effects of climate change since the 1970s are causing about 140,000 additional deaths each year.[18] Extensive research also indicates that over the long term, higher temperatures will increase the levels of ozone and other air pollutants that provoke cardiovascular and respiratory diseases, as well as pollen and other aeroallergens that trigger asthma, affecting the poor and elderly most severely.

Climate change is likely to have the most immediate effect on health and well-being via the declining availability of water. It is estimated that one quarter of the population in Africa (about 200 million people) experience water stress. Increasing temperatures and more variable precipitation are expected to reduce the availability of fresh water, making it more difficult to fulfil basic needs for drinking, cooking and washing. Meanwhile, a greater incidence of floods due to more intense precipitation, sea-level rise and other factors will further contaminate freshwater supplies and increase water scarcity. It will also create opportunities for the breeding of mosquitoes and other disease vectors as people are forced to store water for longer periods.

Higher temperatures and more extreme heat waves are also expected to increase mortality rates, both from the heat itself, and from increased air pollution. Urban areas will be particularly hard-hit due to the "heat island effect", which can make urban areas warmer than surrounding areas, especially at night, by 5° C or more.[19] The summer of 2003 provided a stark reminder of the potentially devastating impacts of heat waves: temperatures that soared above the range of historical experience over large parts of the European continent are estimated to have caused as many as 70,000 deaths, the majority of which occurred in urban areas.

Meeting the challenge of adaptation

Despite the imminent threat, adaptation to climate change has not been mainstreamed into decision-making processes in developing or developed countries. Instead, it tends to be addressed by adding an "extra" layer to existing policy designs and implementation mechanisms. The frequent approach of equating adaptation measures with emergency relief, and framing the challenge in terms of requests for donor support, has not helped. Instead, it has often given rise to a bifurcated approach to adaptation, where efforts either focus on responses to the impacts of climate change (coping measures) or on reducing exposure by climate-proofing existing projects and activities, particularly in the context of disaster risk management. The underlying philosophies can pull in different policy directions. There is a real danger, already apparent in the response to natural disasters, that underlying structural causes of vulnerability and maladaptation will be missed, including a number of closely interlinked and compounding threats to social and economic security.[20]

Recent efforts to forge a more consistent approach to adaptation stress the central role of market incentives. However, this approach also tends to treat the challenge as a series of discrete and unconnected threats which can be addressed through incremental improvements made to existing arrangements, thereby missing the large-scale investments and integrated policy efforts that are likely to be needed.

An alternative is to focus on building resilience to climatic shocks by realizing higher levels of socio-economic development that can provide a buffer for threatened communities. This helps address the interrelated socio-economic vulnerabilities, such as a narrow economic base, limited access to financing, and food insecurity, that can hold back growth prospects and expose communities to unmanageable shocks.

From this perspective, well-designed adaptation measures for addressing climate threats should simultaneously meet other needs, avoid conflict with development objectives, and avoid measures that increase vulnerability to other environmental risks. For example, adaptation to climate change in agriculture should be part of broader agricultural policy efforts to raise productivity and reduce the vulnerability of the sector to outside shocks. Similarly, forest conservation and reforestation policies should be an integral part of broad development and poverty reduction strategies, encompassing investment in economic diversification, human capital and employment creation as well as improvement of land, soil and water management. However, the room for "win-win" (or "no-regrets") solutions should not be exaggerated. The cost of adaptation is likely to be high, and a majority of solutions will involve difficult choices and trade-offs. Making such choices will require enhanced national regulatory authority and strategic planning processes encompassing open discussion within the entire community.

More effective and inclusive institutional responses to the adaptation challenge are also needed. The scale of resources required will, in most cases, call for national resource mobilization and an integrated and strategic approach. Integration of adaptation measures into overall planning and budgeting should start with the assessment of local vulnerabilities to existing climate threats, including their variability and extremes, and the extent to which existing policy and development practice has served to reduce or increase those vulnerabilities. In doing this, policymakers must engage more closely with the affected local communities. They should also draw lessons from past government failures due to insufficient dialogue and cooperation among different ministries, and invest in new capacities to deal with the specifics of the adaptation challenge. For example, meteorological services in many developing countries will need to be improved so they can provide agriculture with more reliable forecasts.

An initial step towards a more integrated approach has been taken by some countries through National Adaptation Programmes of Action, conceived as a means through which least developed countries could secure financial support for climate adaptation. The concept was negotiated during the seventh session of the Conference of the Parties to the United

Nations Framework Convention on Climate Change in Marrakech, Morocco, in 2001. These programmes, which are structured through a bottom-up approach, are action-oriented and tailored to specific national circumstances; they identify "urgent and immediate" investment projects that could significantly contribute to adaptation and poverty alleviation. Some of their greatest strengths are the participation of government agencies and civil society, the consistency with national development plans, and the focus on vulnerability assessment. Still, they are hindered by difficulties in scaling up projects, funding and institutional shortcomings, and a failure to adopt a more broadly developmental approach.[21]

Climate-smart development

In managing climate change, its impacts should be considered in concert with other processes of change, such as urbanization, economic development, and shifts in land use and resource demands. Development policy must become "climate-smart" through its awareness of the range of risks that will emerge over the coming decades. The commitment of resources to meet these risks will be beneficial if it protects the growth path from unforeseen and large-scale shocks, but it does carry a cost, diverting resources from other productive investments. Policymakers must plan adaptation efforts accordingly, with an eye to boosting broader development efforts and special attention to:

- ***Vulnerable populations***, whose "coping range" with respect to climate shocks is limited. For example, some 28 per cent of people in the Mekong and Red River Deltas (some 8.7 million) are small-scale farmers who are estimated to be actually or potentially food-insecure. A changing climate may further stress their livelihoods, with salinity intrusions in the summer and risks of higher-than-historic flooding in the monsoon season. Thus the local impacts of a changing climate would be devastating and would require priority considerations in adaptation plans. This does not mean there should be an exclusive focus on the "poorest of the poor", however; often, for relief purposes, it is better to scale up programmes to include wider rural groups who are at risk for spells of economic insecurity and poverty.
- ***Synergies*** in responding to multiple development challenges. The failure of key infrastructure systems typically results not from a single factor, but from a combination of risks. For example, irrigation could become more difficult due to climate change, exacerbating the impact of

unemployment and food scarcity. The interrelation between adaptation and mitigation also provides opportunities for synergies, for example, where irrigation systems expanded to meet adaptation challenges can be used to open up new markets for low-emissions technologies such as those developed to provide renewable energy.

- ***Scale economies***, resulting from opportunities such as the development of an entire river basin or coastal zone, encompassing major, long-term infrastructure investments in coastal roads, hydropower and irrigation systems. For example, the maritime coast of Mozambique, one of the longest in Africa, extends over a distance of 2,400 kilometres, and is home to about 60 per cent of the population. The area's leading economic activities, including fisheries, tourism and ports, as well as mining, oil and gas, are of immense economic value today, and will continue to be in the future. Ecological and economic stresses already threaten the region, however, and climate change is expected to increase the incidence of destructive cyclones. The Government of Mozambique has drawn up ambitious plans for the sustainable development of the coastal region, including infrastructure (transportation, drainage and water supply), land-use changes, and soft options to manage beach erosion. Such plans, which present unique opportunities for a massive stimulus to development, need to deal with climate risks in an integrated manner, across seasonal, inter-annual and multi-decadal time scales.
- ***Complementarities***, achieved by piggybacking on efforts already under way, such as the expansion of a metropolitan water supply and sewerage system. For example, concerns about maintaining a steady water supply for a hydropower plant on the Rio Amoya in Colombia have led to efforts to preserve the *páramo* ecosystem upstream, in the Las Hermosas massif in the central range of the Andes.[22]

How to apply the integrated approach

To tackle the underlying vulnerabilities that put communities at risk in the face of climate-related threats due to global warming, states must ensure that climate risks are integrated into national and local disaster risk-reduction plans. To be effective, adaptation strategies will have to differentiate among the various dimensions of adaptation at the local, regional, national and international levels as well as within different economic sectors. Table III.1 provides examples of potential adaptation measures for different sectors following the integrated developmental approach suggested above.

Table III.1
Potential measures of adaptation to climate change for different sectors

Sector	*Adaptation measures*
Urban planning	Building residences closer to workplaces in order to reduce transportation time and costs, thereby boosting productivity in a service economy
Water	• Expanded rainwater harvesting • Water storage and conservation techniques • Desalinization • Increased irrigation efficiency
Agriculture	• Adjustment of planting dates and crop diversification • Crop relocation • Improved land management, for example, erosion control and soil protection through tree planting
Infrastructure	• Improved seawalls and storm surge barriers • Creation of wetlands as a buffer against sea-level rise and flooding
Settlement	Relocation
Human health	• Improved climate-sensitive disease surveillance and control • Improved water supply and sanitation services
Tourism	Diversification of tourism attractions and revenues
Transport	• Realignment and relocation of transportation routes • Improved standards and planning for infrastructure in order to cope with warming and damage
Energy	Strengthening of generating facilities and grids against floods, windstorms and heavy precipitation

Source: Adapted from table 5-1 in Dodman, Ayers and Huq, 2009.

Forestry and agriculture

Adaptation practices for the forestry sector today are generally based on lessons learned from past adaptations to climate variability. Important elements of forest protection include improved climate forecasting and disease surveillance systems, and strategies for preventing and combating forest fires, including the construction of fire lines, controlled burning and the utilization of drought- and fire-resistant tree species, such as teak, in tropical forest plantations. Some additional measures aimed at assisting forests in adapting to climate change are also needed to enable sustainable forest management. These include facilitating the adaptive capacity of tree species by maximising silvicultural genetic variation, and management approaches such as minimising slash, reduced-impact logging and widening buffer strips and firebreaks.

In order to succeed in the long run, forest adaptation measures need to include development of alternative, sustainable economic activities for the affected communities. For instance, the Brazilian Amazon is home to 20 million people, many of them poor, and many of them dependent on activities linked to deforestation, such as logging and use of cleared land. This ongoing deforestation releases 200 million tons of carbon into the atmosphere each year, the Brazilian Government reports—about 3 per cent of total global greenhouse gas emissions from all sectors.[23]

In many poorer countries, increasing agricultural productivity and reducing vulnerability to climatic shocks are crucial to long-term sustainability. Decreasing crop failure and maximizing yields over good and bad years alike will be important, particularly where subsistence farming is involved. Strategies to decrease crop failures will include crop diversification, which is potentially one of the most important strategies for achieving food security in a changing climate, and the use of new crop strains that are more weather-resistant and have higher yields. For example, at the Njoro division in Kenya, farmers have been trying to switch from wheat and potatoes to quick-maturing crops such as beans and maize, while planting every time it rains since there is no longer a clear-cut growing season. Still, it is not clear how sustainable this strategy will be, in particular given the multiple vulnerabilities that such communities often face.[24]

In Bangladesh, whereas people traditionally grew low-yield deep-water rice during the monsoon season, they now grow, in areas covered by flood management projects, one high-yield rice crop (*aman*) that is planted during the monsoon, another (*boro*) that is planted in the dry season, with irrigation, and a third (*aus*) that is planted in the pre-monsoon season. Innovative approaches to protecting agriculture in Bangladesh, which is particularly prone to natural hazards and frequent flooding, also include *dap chas* (floating gardens), where crops are grown on floating rafts to protect them from floods.[25]

The interconnections of the risks arising from development and climate are particularly apparent when considering food security. In the Sudan, persistent and widespread drought is likely to worsen with climate change. On the other hand, a more integrated approach to climate risk and livelihoods has increased resilience in some communities. Water harvesting, new crops and types of livestock, and rehabilitation of rangelands, along with access to finance and improving farm skills, have enhanced capacity for adaptation and improved food security.[26]

More generally, economic policies to promote agricultural development should focus on extending support services, particularly for smallholders, and improving infrastructure (such as roads, storage facilities and irrigation networks). Those policies should address land reform and build research and technical capabilities. The establishment of strategic food reserves, including at the international level, would allow governments to reduce price volatility by releasing food in times of emergency and crisis. These reserves could help poor countries which may not have the capacity to respond quickly to sudden scarcity, while proving more effective than other approaches in controlling international price volatility. The need to adapt to climate change could reinforce strategies to promote adaptive agricultural research and development, particularly in Africa, where there is a large gap between current yields and agricultural potential. For example, a new rice variety was successfully developed by the government rice research station in Sierra Leone that has a higher yield and is better suited to drier climate conditions. The new variety is already in use among farmers.

Urban environments

Urban adaptation requires a long-term perspective, recognizing that climate change may amplify the underlying vulnerabilities associated with rapid urbanization. The strain on cities in developing countries is already enormous; adding climate change to the picture may require a paradigm shift in urban planning. National policies to identify and influence formal and informal urban developments are essential, as is the allocation of areas for housing development in order to anticipate and shape the vision for cities. Preventing informal settlements in areas that should not be developed requires governance structures and a solid institutional basis, with city visions and master plans supported by an institutional fabric. Such a fabric is often weak or non-existent in many developing countries.

Disaster risk reduction is an important part of adaptation in urban areas. Institutions established to address disasters are typically weak, and the traditional focus is on disaster relief, consisting largely of searching for missing persons and providing short-term shelter and food. Anticipatory adaptation, in contrast, would encompass preparedness, including relief plans and awareness-raising activities. Thus, looking beyond the emergency dimension of post-disaster response, anticipatory adaptation will need to focus on infrastructure, land-use planning and regulatory measures. Particular attention will need to be paid to temporary dwellings, such as shanty towns and slums, as well as areas built in vulnerable locations and in high-risk areas, such as river banks or unstable hill slopes; in many developing countries sewage and drainage systems need to be built to reduce the risk resulting from more intense precipitation. Some current approaches, such as the elevated walkways set up in Bangkok to cope with flooding, are mere stopgap measures designed to increase pedestrian mobility in high-traffic areas, rather than to shield people from exposure to stagnant surface water.

Long-term measures should address climate vulnerability in the context of rapid urbanization. This would include, for instance, changing legislation that withholds tenure and thus obstructs the consolidation of buildings, contributing to the perpetuation and expansion of shanty towns. Adaptation in urban areas requires strong governance that is focused on sustainable development and supported by appropriate institutional arrangements. As things currently stand, most of the risk to urban areas is in fact associated with the incapacity of local governments to provide for infrastructure, disaster risk reduction and disaster preparedness.

Health and water security

Protection from and adaptation to climate change risks are part of a basic preventive approach to public health, not a separate or competing demand. However, while the global health community has a wealth of relevant experience, shortfalls in provision of basic public health services leave much of the global population exposed to climate-related health risks, making it difficult for health services to look beyond the horizon of urgent short-run health gaps. Thus, there is a need for both additional investment to strengthen key functions, and forward-looking planning for systems that will be able to address new challenges posed by climate change.

Adaptation in the public health realm also requires a broader cross-sectoral approach, as the risks that climate change poses to health are embedded within the wider challenge of achieving genuinely sustainable development. The links between poverty and vulnerability to climate change are clearly in evidence in the health sector, highlighting the importance of further development. Indeed, poverty is one of the leading determinants of vulnerability to climate-related health risks.

Improved water management can have a direct impact on development opportunities, because it is primarily poor water management and lack of water entitlements, rather than physical scarcity of water, that generate water-related tensions and poverty. Along these lines, Bangladesh has begun a pilot project to remove the mountains of silt that build up in its rivers in order to fill in shallow lowlands prone to flooding, or to create new land in order to protect its long, exposed coast against sea-level rise. The silt-trapping experiment has yielded visible gains in small areas such as Beel Bhaina, a low-lying 243-hectare (600-acre) soup bowl of land on the banks of the Hari River, about 55 miles upstream from the Bay of Bengal. United States scientists have recommended a similar silt diversion programme: opening Mississippi River levees south of New Orleans to allow sediment-rich water to flow over the region's marshes, which have been starved of silt since levee-building began in the region hundreds of years ago.

An even greater threat is the increased variability of water supplies in a changing climate. As demand rises due to population growth, increased resilience is required in water management systems. Although efforts are already under way to strengthen these systems in a number of developing countries, significant public investment will be needed to achieve sustainable results.

International cooperation on adaptation

International cooperation on adaptation is essential for a number of reasons. First, the heaviest impact of human-induced climate change will be on small island developing States and the poorest countries in the world, including many African nations. These are the countries that have contributed least to the problem of global warming. Second, these and other developing countries often have difficulties mobilizing the resources needed to reduce their climate exposure, build up resilience and make a rapid recovery after disasters strike. This is a development challenge that can be properly met only through large-scale investments and strategic policies that can draw on the assistance of the international community to help strengthen economic and social capacities at the local and national levels. Third, finding the right response to adaptation can point the way to developing more integrated responses to other shocks that threaten peace, security and well-being.

Setting aside their responsibility for the heightened threats from climate change, the fact remains that developed countries themselves stand to benefit from helping developing countries adapt. The wider consequences of climate impacts, such as increased destabilization and violence resulting from climate-induced conflict, have the potential to jeopardize national and international security. Moreover, the rising level of global inequality that could result from climatic shocks is in neither the economic interest (given the lost export and investment opportunities involved) nor the political interest (given the threat to global cooperation) of rich countries seeking to forge a global framework for better managing climate change. Developing countries, in turn, should give priority to formulating plans for adaptation and take advantage of expertise made available by adaptation funds to establish more integrated and transparent strategies, which would include close consultation with and the participation of their citizens most immediately affected by rising temperatures and climatic shocks.

Scientists confirm that the time frame for acting to curb global greenhouse gas emissions and reduce the probability of catastrophic events is no more than decades—perhaps only the next few decades. Estimates of the cost of adaptation are still tentative and incomplete. The risk, however, lies in underestimating the scale of the challenge, which is made even greater by the slow pace to date of climate change mitigation efforts.

Funding for adaptation has grown substantially in recent years, with almost $4 billion provided in 2009 by the top four bilateral finance institutions alone, according to the United Nations Environment Programme.[27] In addition, there are several specialized funds, including three under the Global Environment Facility and the Adaptation Fund under the Kyoto Protocol. One recent analysis found that six major specialized funds were providing $100 million to $200 million per year for adaptation. It also noted that adaptation funding could increase substantially as a result of commitments made in the Copenhagen Accord and the Cancún Agreements—which include up to $30 billion per year in "Fast-Start" funds for 2010–2012, with "balanced allocation between adaptation and mitigation"—and in the long term, with the new Green Climate Fund.[28] Even so, there is an enormous gap between the resources considered to be necessary for adaptation, in the range of $50 billion–$100 billion per year, and the amount actually mobilized and available.

A key issue with adaptation funding is its relation to development aid. The difficulty in scaling up aid is a real cause for concern, given the urgency of the adaptation challenge in many countries. The current bilateral instruments are unlikely to suffice: more innovative (and predictable) sources of funding will be needed. The principles set out in the UNFCCC, which distinguish between development and adaptation financing, insist on the need for additional funds above existing development aid. This rightly highlights the responsibility of rich countries for funding adaptation, but it runs the risk of ignoring the interconnected nature of these two sets of challenges, of sidestepping the long-standing debate on the adverse impact of excessive policy conditionality attached to aid and other development financing, and of leading to a proliferation of funding mechanisms and facilities that would likely reduce the effectiveness of international support (see chap. VI for further discussion).

Conclusion

All countries will face the challenge of adapting to warmer temperatures and related climatic changes in the coming decades, even if rapid progress is made towards a lower-emissions global economy. However, for some, the threat to livelihoods is already very real and, in some extreme cases, approaches catastrophic levels.

The adjustments required to adapt to climate change cannot be assessed in isolation or undertaken incrementally. Rather, they are closely interconnected with other risks and vulnerabilities that accompany development and will be heavily constrained by local institutional and technological conditions. Successful adaptation hinges on faster and more equitable growth, even as failure to adapt threatens those goals.

This chapter has argued that, in many cases, the response will involve a sizeable investment of resources to make countries and communities more resilient and to address vulnerabilities that can turn even small climatic shocks into long-term development disasters. This excludes a one-size-fits-all policy response. The right approach is an integrated national strategy, which will require mobilization of domestic resources and the guidance of an effective developmental state.

Meeting such challenges will require a break with recent policy approaches which have given undue attention to market forces and competition. Adaptation, like mitigation, is a public policy challenge, the complexity of which will require using a broad array of strategies to build resilience.

A smarter approach will build adaptation responses into ongoing development strategies by paying particular attention to vulnerable populations, by making use of large public works and taking advantage of scale economies, by addressing the thresholds below which current systems consistently fail, and by exploiting investment complementarities.

Even so, many countries for whom the challenge is simply too big cannot be expected to meet it by themselves. Hence, it was agreed at the UN Climate Change Conference in Bali in 2007 that finance and technical assistance would be available to help developing countries meet the adaptation challenge. Years later, that assistance remains woefully inadequate and poorly organized. Improvements in this regard are likely to be a prerequisite for making real headway towards putting those countries on more sustainable development paths.

Notes

1 Hansen, J., Sato, M., Kharecha, P., Beerling, D., Berner, R., Masson-Delmotte, V., Pagani, M., Raymo, M., Royer, D. L. and Zachos, J. C., 2008. 'Target Atmospheric CO2: Where Should Humanity Aim?' *The Open Atmospheric Science Journal*, 2. 217–31. doi:10.2174/1874282300802010217.

2 Betts, R. A., Collins, M., Hemming, D. L., Jones, C. D., Lowe, J. A. and Sanderson, M. G., 2011. 'When could global warming reach 4°C?' *Philosophical Transactions of the Royal Society A: Mathematical, Physical and Engineering Sciences*, 369(1934). 67–84. doi:10.1098/rsta.2010.0292.

3 United Nations Development Programme, 2007. *Fighting Climate Change: Human Solidarity in a Divided World*. Human Development Report 2007/8. New York. http://hdr.undp.org/en/reports/global/hdr2007–8/.

4 United Nations Framework Convention on Climate Change, 2007. *Climate Change: Impacts, Vulnerability and Adaptation in Developing Countries*. http://unfccc.int/resource/docs/publications/impacts.pdf

5 See, for example, Loughry, M., and McAdam, J., 2008. 'Kiribati: relocation and adaptation.' *Forced Migration Review*, 31. http://www.fmreview.org/FMRpdfs/FMR31/51–52.pdf. 51–52.

6 Epstein, P. R. and Ferber, D., 2011. *Changing Planet, Changing Health: How the Climate Crisis Threatens Our Health and What We Can Do About It*. University of California Press, Berkeley. 3.

7 Moore, S., 2011. 'Strategic imperative? Reading China's climate policy in terms of core interests.' *Global Change, Peace & Security*, 23(2). 147–57. doi:10.1080/14781158.2011.580956.

8 United Nations Human Settlements Programme (UNHabitat), 2008. *State of the World's Cities 2008/2009: Harmonious Cities*. Earthscan, London.

9 United Nations, 2006. *World Economic and Social Survey 2006: Diverging Growth and Development*. New York. http://www.un.org/en/development/desa/policy/wess/wess_archive/2006wess.pdf

10 United Nations Conference on Trade and Development, 2002. *Economic Development in Africa—From Adjustment to Poverty Reduction: What Is New?* UNCTAD/GDS/AFRICA/2. Geneva. http://www.unctad.org/templates/WebFlyer.asp?intItemID=2868&lang=1.

11 Intergovernmental Panel on Climate Change, 2007. *Climate Change 2007: Impacts, Adaptation and Vulnerability. Contribution of Working Group II to the Fourth Assessment Report of the Intergovernmental Panel on Climate Change*. M. L. Parry, O. F. Canziani, J. P. Palutikof, P. J. van der Linden, and C. E. Hanson (eds.). Cambridge University Press, Cambridge, UK.

12 Cline, W. R., 2007. *Global Warming and Agriculture: Impact Estimates by Country*. Center for Global Development & Peterson Institute for International Economics, Washington, D.C.

13 Schlenker, W., Hanemann, W. M. and Fisher, A. C., 2005. 'Will U.S. Agriculture Really Benefit from Global Warming? Accounting for Irrigation in the Hedonic Approach.' *The American Economic Review*, 88(1). 113–25. doi:10.1257/0002828053828455.

14 Schlenker, W. and Lobell, D. B., 2010. 'Robust negative impacts of climate change on African agriculture.' *Environmental Research Letters*, 5(1). 014010. doi:10.1088/1748-9326/5/1/014010.

15 Butt, T. A., McCarl, B. A., Angerer, J., Dyke, P. T. and Stuth, J. W., 2005. 'The economic and food security implications of climate change in Mali.' *Climatic Change*, 68(3). 355–78. doi:10.1007/s10584-005-6014-0.

16 Food and Agriculture Organization of the United Nations, 2004. *Trade and sustainable forest management: impacts and interactions*. Analytic study of the global project GCP/INT/775/JPN. Impact assessment of forests products trade in the promotion of sustainable forest management. FAO Forestry Department, Rome.

17 United Nations Human Settlements Programme (UNHabitat), 2007. *Global Report on Human Settlements 2007: Enhancing Urban Safety and Security*. Earthscan, London.

18 World Health Organization, 2010. *Climate and Health*. Fact sheet N°266. Geneva. http://www.who.int/mediacentre/factsheets/fs266/en/.

19 See, for example, Aniello, C., Morgan, K., Busbey, A. and Newland, L., 1995. 'Mapping micro-urban heat islands using LANDSAT TM and a GIS.' *Computers & Geosciences*, 21(8). 965–67. doi:10.1016/0098-3004(95)00033-5.

See also Patz, J. A., Campbell-Lendrum, D., Holloway, T. and Foley, J. A., 2005. 'Impact of regional climate change on human health.' *Nature*, 438(7066). 310–17. doi:10.1038/nature04188.

20 See also chapter III in Vos, Rob and Kozul-Wright, Richard, eds., 2010. *Economic Insecurity and Development*, New York: United Nations (Sales E.11.II.C.3).

21 For more on NAPAs, see United Nations Framework Convention on Climate Change, n.d. 'National Adaptation Programmes of Action (NAPAs).' http://unfccc.int/national_reports/napa/items/2719.php. [Accessed 29 July, 2011].

For a critique of NAPAs, see Huq, S., and Osman-Elasha, B., 2009. 'The status of the LDCF and NAPAs.' Presentation at the International Scientific Congress on Climate Change: Climate Change: Global Risks, Challenges and Decisions (Copenhagen, 10–12 March 2009), session 41, *Adaptation to climate change in least developed countries: challenges, experiences and ways forward*. http://www.ddrn.dk/filer/forum/File/IARU_2009_Session_41_SaleemulHuq.pdf

22 The World Bank, Carbon Finance Unit, n.d. 'Colombia: Rio Amoya Run-of-River Hydro Project.' http://wbcarbonfinance.org/Router.cfm?Page=Projport&ProjID=54401. [Accessed 29 July, 2011].

23 Azevedo-Ramos, C., 2007. *Sustainable Development and Challenging Deforestation in the Brazilian Amazon: The Good, the Bad and the Ugly*. Brazilian Forest Service, Ministry of Environment, Brasilia. http://www.fao.org/docrep/011/i0440e/i0440e03.htm.

24 Dodman, D., Ayers, J. and Huq, S., 2009. 'Building resilience.' In The Worldwatch Institute, *State of the World 2009: Into a Warming World*. W.W. Norton and Company, New York. http://www.worldwatch.org/files/pdf/SOW09_chap5.pdf

25 Banerjee, L., 2007. 'Effect of Flood on Agricultural Wages in Bangladesh: An Empirical Analysis.' *World Development*, 35(11). 1989–2009. doi:16/j.worlddev.2006.11.010.

26 Osman-Elasha, B., Goutbi, N., Spanger-Siegfried, E., Dougherty, B., Hanafi, A., Zakieldeen, S., Sanjak, E., Atti, H.A. and Elhassan, H.M., 2008. 'Community development and coping with drought in rural Sudan.' In *Climate Change and Adaptation*, N. Leary, J. Adejuwon, V. Barros, I. Burton, J. Kulkarni and R. Lasco (eds.). Earthscan, London.

27 United Nations Environment Programme, 2010. *Bilateral Finance Institutions and Climate Change: A Mapping of 2009 Climate Financial Flows to Developing Countries.* Report prepared by the Stockholm Environment Institute and the UNEP Climate Change Working Group for Bilateral Finance Institutions. http://www.unep.org/pdf/dtie/BilateralFinanceInstitutionsCC.pdf

28 Smith, J. B., Dickinson, T., Donahue, J. D. B., Burton, I., Haites, E., Klein, R. J. T. and Patwardhan, A., 2011. 'Development and climate change adaptation funding: Coordination and integration.' *Climate Policy*, 11(3). 987. doi:10.1080/14693062.2011.582385.

Chapter IV
A state of change: climate and development policy

Introduction

The previous chapters have suggested that there are climate-friendly alternatives for development that steer clear of carbon-intensive technologies. This chapter considers the national policies that might be necessary to support what amounts to a new industrial revolution in developing countries.

Economic and technological revolutions in the last two centuries have opened up opportunities for "latecomers" to kick-start a process of rapid growth and development. However, not all were able, or allowed, to seize those opportunities. Meanwhile, the economic gains to "first movers" have often been cumulative, resulting in divergent economic development patterns and rising gaps in incomes, technological capacity and energy use.

Those precedents are a concern for developing countries, which fear being priced out of new low-carbon technologies while being asked to forgo the cheaper, older technologies that others have relied on in the past. Moreover, the latest technological revolution is unfolding at a time of profound global economic and financial stress that makes it even harder for poor and vulnerable countries and communities to shift to a new economic and technological paradigm.

The Commission on Growth and Development, an international collaboration that sought to identify policies and strategies to support economic growth and poverty reduction, argued that a conceptual impasse has been reached in the debate on how to cut carbon emissions while allowing developing countries to grow.[1] To remove that impasse, this chapter argues for a big push, blending pro-investment macroeconomic and industrial policies in order to move to a transformative low-emissions growth path. The integrated development strategy needed to achieve this, in turn, requires a strong and dynamic developmental state to manage it and sufficient policy space to be able to adapt climate measures to local needs and sensitivities.

The following sections look at how the traditional functions of the developmental state relate to the climate challenge, the role of industrial policy in an investment-led strategy for meeting climate and development goals, and some specific measures that could help policymakers in developing countries begin the transition to a low-emissions, high-growth strategy.

The developmental state in a warming world

An investment-led strategy

All the well-known stories of successful economic development have involved a sustained burst of growth that raised overall incomes and living standards significantly. When countries follow an inclusive development path, growth is accompanied by poverty reduction and improvements on a broad set of social indicators. This path does not emerge spontaneously, however, and even after a period of rapid growth, countries can get stuck or fall back.

A sustained acceleration of growth usually requires rapid capital accumulation and shifts in the structure of economic activity towards high-productivity sectors. An important part of the early development policy debate focused on how to quickly raise the share of investment in national income to a level that would trigger a virtuous circle of rising productivity, increasing wages, technological upgrading and social improvements. The required investments are often closely connected, depend on reaching a minimum scale to be efficient, and may become profitable only over a long time period. The presence of scale economies, complementarities, threshold effects and externalities limits the role that market forces by themselves can play in realizing the desired investment path. Infrastructure development in general and energy supply in particular have always been critical elements in this story and, as discussed in previous chapters, the importance of the latter has grown in the context of climate change.

Focusing the "big push" on selected leading sectors can attract further investment through the dynamic effect of decreasing costs and the expansion of strong backward and forward linkages. In this regard, the strategy is less about detailed planning and more about strategic support and coordination, with a significant role for public investment in triggering growth and crowding in complementary private investment. A given rate of capital accumulation can, of course, generate different rates of output growth, depending on its nature and composition and on the efficiency with which production capacity is utilised. Policies will have a significant bearing on the outcome. Over the years, the minimum scale of investments needed to launch and maintain an industrialization drive has risen steadily.

In most success stories, a developmental state has helped promote long-term growth and structural change by increasing the supply of investible resources and socializing long-term investment risk. Along with targeting resources into high-productivity activities, states have provided predictable and affordable credit through a managed financial system, adopted pro-investment macroeconomic policies, and directly invested in some key sectors. The East Asian economies have often been held up as exemplary embodiments of the developmental state, but there are many other such examples (for an example from United States history, see box IV.1).

Box IV.1: The Tennessee Valley Authority: A successful big push

The post-war economic boom in the American South, which followed large public capital investments during the New Deal and the Second World War, is a successful example of a big push. By triggering an increase in the rates of return to private investment, the infusion of public capital through the Tennessee Valley Authority (TVA) provided a major impetus for the rapid post-war industrialization of the Southern economy. (The focus here is exclusively on the historical role of the TVA, not on its current operations and policies.)

The TVA was established in 1933 as part of the New Deal, intended by President Franklin D. Roosevelt to lift the United States out of the depths of the Great Depression. It was conceived both as a development agency, mandated to raise living standards in the Tennessee River Valley, and as a construction and management agency mandated to build and operate dams and structures along the Tennessee River, whose drainage basin over seven states covers some 40,900 square miles (or 105,900 square kilometres). The TVA was to function as, in Roosevelt's words, "a corporation clothed with the power of government but possessed of the flexibility and initiative of a private enterprise."

Over the 12-year period from its inception in 1933 to the end of the Second World War in 1945, the TVA established its institutional framework, built broad-based local support for its programmes, and developed a physical infrastructure that would serve as the backbone for its accomplishments. This infrastructure included a vast system of multi-purpose dams and reservoirs designed to harness the potential of the Tennessee River and an extensive transmission system created to provide cheap electricity throughout the region. Early and intense efforts to improve agriculture, land use and forestry practices helped to restore and maintain a healthy environmental base, while access to small-scale credit and technical assistance programmes provided the citizens of the Valley with the tools they needed to improve their own lives. It was during those early years that the Tennessee Valley Authority established what may have become its greatest legacy: the integration of a healthy natural resource base, a strong infrastructure, and human capacity to foster the social and economic development of a region.

The need for TVA arose from the dire social and economic conditions in the Tennessee Valley in the 1930s. As the Great Depression of the 1930s deepened

Box IV.1 (cont'd)

and conditions in the Tennessee Valley (and elsewhere in the nation) worsened, Roosevelt sought to create an innovative programme that would revitalize the economy and boost morale. The creation of TVA represented a "bold experiment" aimed at accomplishing the unified development of a river basin. Flood control, navigation and power generation were not ends in themselves, but the means to advance social and economic development.

Although rich in natural resources, the region was largely rural and undeveloped, poverty-stricken and characterized by degraded environmental conditions. Per capita income was among the lowest in the United States, few people had running water or electricity, and poor sanitary conditions resulted in some of the highest rates of disease and infant mortality in the country. In some areas near the Tennessee River, 1 out of every 3 people had malaria. Illiteracy rates were high and the quality of education was poor. Severe erosion, extensive deforestation and exhausted mines were indicative of a deteriorating environment. Additionally, the navigation potential of the Tennessee River remained untapped owing to hazardous shoals, while the heavy rainfall in the region subjected many areas to repeated and serious flooding. The people of the Tennessee Valley were trapped in a cycle of poverty. The social problems in the Valley could be addressed only by improving the economy, which would depend on a healthy resource base, including land, water and forests.

The vitality of the TVA as an institution was bolstered by its early, tangible and largely positive impact on the lives of the people of the Tennessee Valley. Two major dam construction projects were initiated in the agency's first year of operation. Over the next 12 years, bolstered by the need to support the war effort, progress was remarkable: the navigation channel on the Tennessee River was completed; 26 dams were incorporated into the TVA water control system; and TVA became the largest power producer in the United States. Additionally, farm production levels tripled owing to successful efforts to reduce soil erosion, improve farm practices and introduce fertilizers. Although controversies arose over relocations required during dam building, the Valley residents were put back to work and the overall standard of living improved. TVA won the support of citizens and local governments and gained a national reputation for its work in the area of water resources, land management, forestry, agriculture and energy production.

Sources: Bateman et al., 2009, and Miller and Reidinger, 1998.

Some developed countries built this kind of investment-led approach to the climate challenge into the stimulus packages they adopted in response to the economic downturn, with investments designed to create "green jobs". Developing countries, however, need to aim for something much bigger, effectively a new industrial revolution that relies on low-emissions and, in due course, carbon-free energy sources. This new investment path, which should involve a broad range of sectors and regions, would weaken the climate constraint on economic growth. Related investments will be needed to raise agricultural productivity, improve forest management, and ensure a more reliable water supply and a more efficient transport system—all of which will entail the steady expansion of green jobs.

From technological learning to leapfrogging

While economic growth depends on a fast pace of investment accumulation, it is sustained by ongoing structural and technological changes which underpin productivity and income growth. Without constant innovation and learning, the economy will remain locked into production methods that use less advanced technology and will fail to diversify into more dynamic activities. Since improved technological knowledge is often embodied in capital goods, rapid capital formation and technological progress are strongly complementary. A pro-investment macroeconomic policy is therefore necessary to strengthen technological development, but private firms will still tend to under-invest in technological knowledge and innovation. Thus catch-up growth requires a good deal of active policy support for building technological capacity, including importing technologies and learning to use them.

Technologies usually evolve with institutions that support them, so there is a tendency to favour (or "lock in") incumbents and "lock out" new technologies. States can help by removing regulatory and institutional barriers that generally favour incumbent technologies. They can also directly support new technologies through their procurement policies and through subsidies, and can provide temporary support to those adversely impacted by the resulting shifts in activity.

Governments have also often provided support for tertiary education, publicly funded research, development and deployment (RD&D) and subsidized research undertaken in the private sector, as well as industry-level training. In recent years, such efforts have focused on establishing national systems of innovation, including stronger partnerships between public and private institutions promoting technological development; however, serious financial and institutional obstacles to building such systems have been identified in many developing countries.[2]

Given the scope of the challenge, the innovation and learning that are critical to a new low-emissions growth path should involve not just the most advanced economic sectors, but also traditional sectors such as agriculture and forestry. Information and communications technologies should play a central role, with their vast potential to support the smart and efficient production, distribution and use of energy; those technologies also offer many organizational, marketing and research-oriented capabilities which will be useful in fostering productivity growth and finding new markets.

An important concept in sustainable energy development is leapfrogging: that is, the hope that developing countries can avoid the traditional, resource-intensive pattern of economic and energy development by "leapfrogging" to the most advanced technologies available, rather than retracing the steps taken by already-industrialized countries. The idea has gained ground among policymakers, researchers, and to some extent, in the private sector.

However, while leapfrogging has the potential to yield important savings over the long run, it faces significant obstacles. On the supply side, there may be barriers to accessing the required technology, whether because of the widespread obstacles to importing it (see chap. V), or because of a lack of expertise needed to link technology to local conditions. Obstacles may also exist on the demand side, if a limited market size prevents economies of scale that would make new technologies locally competitive within an acceptable time frame. Thus, there is a role for governments to build markets for new technologies, for example, by providing low-cost loans to households and businesses, and distributing information about new innovations.

Still, as noted in chap. III, there is an urgent need for a significant scaling up of adaptation capacity. In order to take advantage of the new opportunities, it will be necessary to invest in training institutes and schools and expand the availability of basic education, as well as vocational and technical training. Information technologies also open up new possibilities for remote training.

The "hardware" side of training, or training in core technical skills, may be more important for least developed countries that need to reach the threshold of a skilled labour force in order to be able to absorb technology. In contrast, higher-income developing countries may be more in need of "software" skills, including in business promotion and networking. For small economies, such as small island developing States, regional cooperation can be crucial as a means to achieve economies of scale.

Managing creative destruction

Development is a continuing process of adjustment and transformation. Changes in the economic system require new incentives and regulations to ensure that adjustments are smooth. Institutional change is also needed to ensure open discussion, consultation and broad participation, and to prevent those who stand to lose from thwarting the process. Success rests, to a great extent, on the developmental state's capacity to provide a coherent vision of the future and manage the challenges engendered by change, including overcoming vested interests and supporting those who are losing out.

Phasing out "dirty" technologies, in turn, will not only require finding substitutes, but also avoiding the installation of new, dirty facilities with high sunk costs that lock industries and countries into high-emitting technologies for years to come. Managing such adjustments will be critical to achieving the smooth transition to a low-emissions, high-growth pathway.

These are non-marginal changes that are unlikely to emerge from the play of market forces alone. Old technologies are still cheaper, and we can expect their price to remain relatively low for the foreseeable future. Old technologies are also readily available for replication and installation. While some green technologies are already cost-competitive, others remain expensive, and still others need to be developed.

Governments can fundamentally shape energy demand through land-use, urban and regional planning—that is, through careful spatial planning of different types of economic activities to minimize demand for energy, maximise opportunities for cogeneration, and allow for the efficient development of mass transit systems as well as non-motorized forms of transport.

Thus, tackling climate change requires a strong set of legislative and regulatory incentives to prevent the players from becoming sidetracked by or locked into carbon-intensive options. This will involve significant coordination among different spheres of government. It also means that an integrated development strategy must cover energy, natural resource use, the energy- and natural resource-intensity of production, and a vision of urban development and transportation. The strategy should be launched collaboratively by the state and the private sector.

Diversification challenges

As discussed in the previous chapter, for many developing countries, adapting to unavoidable shocks from global warming is a major policy challenge. While adaptation needs are multifaceted, the core challenge for many countries is to diversify their economies, to reduce their dependence on a few (often climate-sensitive, resource-based) industries, and to shift to new energy sources and to less energy-intensive sectors (see box IV.2).

Box IV.2: Diversification of the productive system in South Africa

Historically, low electricity prices have been seen as central to South Africa's competitiveness. The use of cheap and abundant coal has provided relatively low-cost electricity, leaving little incentive for greater energy efficiency. Industrial development has, to a significant extent, been built around energy-intensive sectors. These sectors are sensitive to changes in energy prices, so that particular attention needs to be given to them in the move to a low-emissions economy. While current government policy has embraced sustainable development goals, the country continues to provide significant incentives for investment in energy-intensive industries. These industries are still an important source of employment, investment and income.

Continuing this approach carries a high risk that the economy will be "locked into" energy-intensive industries, when environmental, economic and social pressures may push South Africa in the opposite direction. Significant investment in energy-intensive industries in the 1990s has had just that effect, and new megaprojects continue to be planned.

An active industrial policy is required to target sectors that are less energy-intensive and enable South Africa's economy to diversify, move away from the country's mineral-energy complex and shift to capital and intermediate goods. This would represent a major shift and could take decades to complete. However, given the lock-in effect, decisions taken today will be critical in changing the trajectory of South Africa's energy development path. "Bending the curve" requires a long-term perspective, but it also involves policy changes in the immediate future.

The most effective and affordable short-term strategy for reducing greenhouse gases emissions in South Africa is an energy-efficiency programme. The next strategy would be to change the fuel mix to reduce the share of coal in the total primary energy supply. In the medium term, reduced-carbon and non-carbon energy supplies, such as natural gas, hydroelectricity (imported from the region) and solar thermal technologies could be introduced. These measures can together achieve significant emissions reductions, but further action will also be required, possibly with the help of international funding.

Renewable energy options in South Africa have been considered both in terms of electricity-generating renewable technologies (a combination of biomass, solar thermal technologies and wind energy) and a biofuels industry. Investing in more labour-intensive technologies such as renewables would create more "green jobs". Other, more ambitious renewable energy interventions are possible, particularly one involving a massive effort to develop solar energy technologies, since South Africa has excellent solar resources, but this would again depend on the electricity price. Current evidence indicates that solar water heating (for domestic, commercial and potentially industrial applications) is economically viable, even given current low prices. Developing the potential of solar energy in South Africa would probably require a massive state-driven research project and an investment programme similar to the synthetic fuels programme of the 1960s and 1970s.

Other supply-side options that require further investigation include new coal technologies and unconventional coal technologies such as fluidized-bed

Box IV.2 (cont'd)

combustion, as well as carbon capture and storage combined with coal gasification. There are currently no reliable estimates for the cost of these programmes, especially given the lack of oil or gas wells in South Africa, a factor that introduces significant technical complications with respect to CO_2 storage. There are also plans to develop a biofuels industry, but on a relatively small scale, limited by cost and the availability of land and water.

To achieve the desired transformation, South Africa will need to promote less emissions-intensive industries and sectors, while tackling the challenge posed by the energy-intensive sectors through a combination of reviewing the existing policy framework, promoting specific energy-intensity targets, conducting international negotiations on the best location for such industries, and carrying out diversification within these sectors. The aim of these strategies would be to protect South Africa's competitive advantage in the short and medium terms, while building other competitive advantages in the long term.

Source: Winkler and Marquand, 2009.

Agriculture, as one of the most climate-sensitive sectors in many countries, will require smart new policies that integrate adaptation and mitigation. One key strategy is to build knowledge of new technologies such as sustainable irrigation methods and crop selection and diversification. A proactive approach can prevent production losses and a further aggravation of the food crisis and poverty in rural areas, especially in Africa.

At present, agriculture is the main emitter of nitrous oxides and methane (both with high global warming potential), contributing about 14 per cent of global greenhouse gas emissions in 2005, more than the transportation sector.[3] At the same time, agriculture is an area with large and relatively inexpensive mitigation potential. One study estimated that by 2030, agricultural emissions could be more than halved from business-as-usual levels through a combination of measures that would cost less than $10 per ton of CO_2 equivalent (CO_2e) abated; many measures would have negative costs because of productivity benefits.[4] Low-cost measures include improving soil quality (for example, restoring degraded lands) and cropland and grazing land management (for example, reducing fertilizer use, reducing tillage and eliminating burning of crop residues in the field).[5] Thus, sustainable agriculture can contribute to meeting climate change mitigation goals as well as the Millennium Development Goals for reducing income poverty and hunger. However, taking advantage of the mitigation and carbon sink potential in agriculture will require capacity-building programmes, with investments in technical training, provision of extension services, and programmes for sharing good practices.

The production of biofuels from biomass is another potential means of mitigating climate change and generating income in the agricultural sector. However, this will require further research on sustainable production methods and management of the potential competition between biofuel and food production, along with extensive farmer and farm worker training. If the biofuel industry grows, it will require not only a large unskilled labour force, but also skilled labour. Training will be needed in the technical and managerial skills needed in the nascent biofuel processing industries.

The appropriate diversification strategy for each country will be context-specific. It depends, among other factors, on the level of development, technological capacities, size of the economy, natural resource base, government capacities and established state-business relations. It is important to remember that a development strategy extends beyond manufacturing, and can include new uses of natural resources and the development of modern services.

The revival of industrial policy

There is growing recognition that there is no "one size fits all" policy approach to development—especially when it is combined with the climate challenge. For many years, developing countries were pressed to adopt a narrow band of generic, market-friendly measures, and to roll back the state. Unfortunately, those policies have seriously debilitated public sector capacity in some countries and left an institutional hiatus which needs to be filled.[6]

Governments have a long history of improving the efficiency of the market system by correcting for market failure, especially in non-competitive markets, and of accelerating growth by providing missing inputs and promoting public-private collaboration in long-term investment, research and development, education and training, etc. Still, government is no less fallible than markets, and unpredictable government actions can also hinder long-term investment. More secure property rights can help address the latter; in addition, stronger and more reliable civil service capacities and public institutions will be needed.

The large up-front investments needed in both mitigation and adaptation will require a coordinated effort to mobilize domestic and external resources and to channel them into high-productivity, energy-efficient activities. Macroeconomic policy instruments would need to be deployed to support a development mandate dedicated to productive investment, structural change and rapid growth. In order for this to be possible, financial markets must be regulated in a manner that protects the autonomy of national governments.[7]

Fiscal and monetary policies should give priority to increasing public spending, including investments in renewable energy, cleaner energy processes, education, health and infrastructure. This will also entail the use of subsidized credit, credit guarantees, tax breaks, accelerated depreciation allowances, and other policy measures to boost profits in private firms in the desired sectors. The effects of such policies will be greater if commercial banks make loans more easily available for such investments. However, as discussed in chap. VI, development banks may have a larger role to play in some countries.

Because a big investment push is likely to target a limited range of industries and sectors and involve substantial public investment, there has been much warning about potentially crowding out private investment. In such scenarios, additional government spending has little positive, or even a negative effect on total output, because of its adverse effects on interest rate-sensitive components of private expenditure. Neither theory nor empirical evidence provides a basis for clear-cut conclusions in these respects, however, and our own big-push scenario allows for considerable crowding in (see chap. I).

Yet pro-investment macroeconomic policies are not enough. There is also a need for effective industrial policies, with some key ingredients: targeted incentives, regulation, coordination of investment decisions, and control mechanisms. These elements can be implemented through diverse instruments, according to the particular characteristics of the sector and country.

In many developing countries, such measures have been used to attract foreign direct investment. These countries have the experience and instruments required to target and tailor industrial policies towards a big push in clean energy and towards diversification in support of greater economic resilience. Some have been more successful than others, however, in using these policies. The subsidies and rents created by these measures need to be conditioned on enhanced performance, linked, for example, to technological upgrading, and limits must be set on how long they can be used. These and other lessons will need to be absorbed as industrial policies are implemented to meet the climate challenge. Of course, the first step is to shift out of reverse and into forward gear: currently, many countries still have policies favouring high-emissions sectors such as hydrocarbons.

A logical, though not easy, starting point is to reorient support from these sectors towards renewable or cleaner energy sources.

Developing countries operate today in a global policy environment that is much changed from two or three decades ago. In particular, there has been a tendency to discipline national economic policies through multilateral, regional or bilateral agreements that restrict countries' ability to adopt certain types of industrial policies. For example, direct export subsidies are now illegal for all but least developed countries, as are domestic content requirements on enterprises that are linked to trade, quantitative restrictions on imports, and patent laws that fall short of international standards. However, there remains much scope for coherent industrial policies, especially if countries do not further erode their policy autonomy.

The ethanol industry in Brazil demonstrates how critical government support can be, particularly in the early phases of a new technology, and how support may need to be sustained until the technology has taken firm root in the marketplace (box IV.3). The government of Brazil, at both

Box IV.3: Brazil's sugar cane-based ethanol industry

Brazil's ethanol industry was established in the 1930s. With more sugar than it could use, the government determined that sugar cane should be utilised for ethanol production and required that ethanol be added to gasoline. Following the international oil crisis in 1973, the industry expanded in importance. The government launched the National Alcohol Programme (Pro-Álcool) in 1975 to increase production yields, modernize and expand distilleries, and establish new production plants. Although ethanol production was initially subsidized heavily, over time the subsidies have been eliminated.

Policies that were key to Brazil's success in promoting ethanol use include (a) obligating the state-owned oil company, Petrobras, to purchase a guaranteed amount of ethanol; (b) providing economic incentives to agro-industrial enterprises in the early 1980s to produce ethanol, including loans with low, subsidized interest rates; (c) encouraging consumers by guaranteeing a price of ethanol at the pump at 59 per cent of the price of gasoline; (d) requiring the automobile industry to produce cars able to run partially or totally on biofuels; (e) allowing renewable energy-based independent producers of electric power to compete with traditional public utility firms in the electricity market at large; (f) stimulating and supporting private ownership of sugar mills, which helped increase competition and efficiency; and (g) stimulating rural activities based on biomass energy to increase employment in rural areas.

The Sugar Cane Technology Centre, a privately funded research institute in São Paulo, played a central role in improving ethanol production technology, investing about $20 million per year in research at the peak of the programme. Researchers at the Centre and other institutions also found ways to use sugar cane fibre residue,

Box IV.3 (cont'd)

known as bagasse, to produce energy, building on existing methods of burning bagasse to power steam turbines for electricity generation and using the remaining heat from the turbines for the distillation process. They developed cauldrons operating at greater pressure so that more energy could be produced, allowing many ethanol plants to become self-powered. This contributed significantly to keeping ethanol production costs low.

Thanks to steady productivity improvements, the cost of producing ethanol declined by an annual average of 3.8 per cent from 1980 to 1985 and of 5.7 per cent from 1985 to 2005. As cumulative experience increased, the cost per unit of energy declined (see figure).

Producers' price of ethanol compared with gasoline prices in Brazil

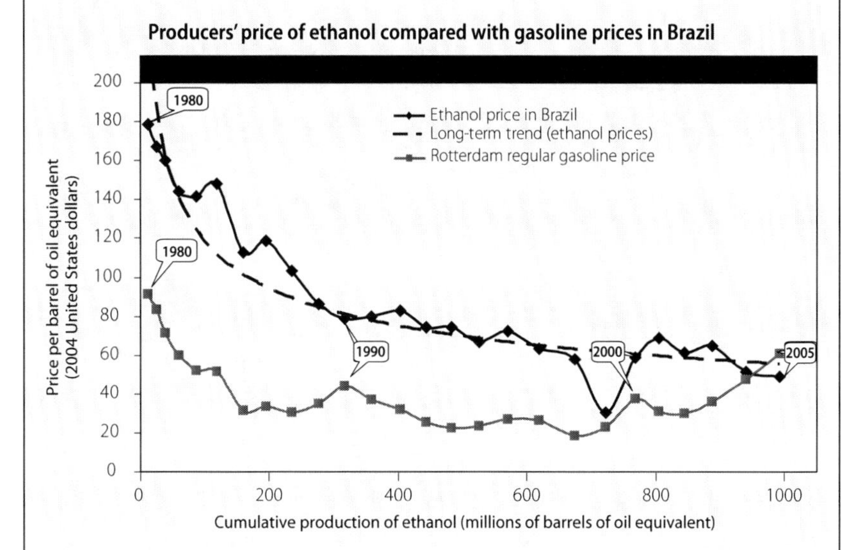

In 2009, Brazil was the second biggest producer of ethanol in the world (25 billion litres) after the United States (38 billion litres), according to the Renewable Fuels Association. Close to 80 per cent of Brazil's ethanol is for the domestic market, where much of the demand has been driven by the success of flex-fuel vehicles. Introduced in 2003, such vehicles can run on gasoline, ethanol or a mix of both; in March 2010, they surpassed 10 million units. According to Brazil's National Association of Automotive Vehicle Manufacturers (ANFAVEA), 86 per cent of cars sold in 2010—almost 3 million—were flex-fuel, and since 2009, flex-fuel motorcycles have been sold in Brazil as well, quickly gaining popularity.

Brazil is also experimenting with another biofuel made from sugarcane, a biodiesel. In 2011, after a successful field test in São Paulo of buses fueled in part by sugarcane diesel, a contract was announced for 160 buses in the city to run on a 10 per cent sugarcane diesel blend through 2012.

The government's reasons for supporting biofuels have expanded to include concerns about greenhouse gas emissions and climate change, rural employment

Box IV.3 (cont'd)

and equity issues, and local air pollution. The use of ethanol as a replacement for gasoline—in flex-fuel vehicles and blended into all conventional fuel—has led to an overall reduction of 9.2 million tons of carbon per year in carbon emissions in Brazil (10 per cent of the total). When used as an additive, ethanol also displaces highly toxic and volatile components of gasoline (such as lead, benzene, toluene and xylene).

Brazil now offers its expertise to other countries, especially developing countries that could produce biofuels but still depend on oil. However, Brazil's own success has been marred recently by supply shortages and a surge in sugar prices since 2010 that significantly increased the cost of Brazilian ethanol and led the country to start importing U.S. ethanol. To reduce the need for imports and subdue inflation, the government lowered the ethanol mandatory blending rate for gasoline from 25 per cent to 18 per cent. In addition, the state-run development bank BNDES announced in June 2011 that it would provide 30 billion to 35 billion reais ($19 billion to $22 billion) to finance expansion in the sugar cane sector through 2014.

Sources: Goldemberg, 2008; Goldemberg et al., 2004; Moreira, 2006; Almeida, 2007; Food and Agriculture Organization of the United Nations, 2008; Nakicenovic, 2009; United Nations, 2008; Institute for International Trade Negotiations of Brazil (ICONE); Brazilian Sugar Cane Industry Association (UNICA), 2010; Riveras and Winters, 2011; Amyris Brasil 2011.

federal and state levels, had an essential role to play in providing incentives to scale up production and in setting up a clear institutional framework. This role included setting technical standards, supporting the technologies involved in ethanol production and use, providing financial advantages, and ensuring appropriate market conditions.

Replacing old technologies, like gasoline in the case of Brazil, with renewable sources requires complementary investments along the supply chain. In the particular case of gasoline, consumers are reluctant to buy cars using a new fuel that may be difficult to find. Service station owners are not interested in investing in a parallel fuel distribution system, since the number of potential users is initially very small. This is why government policies to spur investment and drive demand for selected technologies are so important.

Additionally, in most countries, government is the single largest consumer. Thus, government procurement policies, including methods such as tendering and holding of reverse auctions, can be an important instrument. As major purchasers of electricity and vehicles, governments can give a significant boost to low-emissions options through procurement bidding specifications. Such green procurement could also extend to new construction of government buildings, ranging from offices to schools and hospitals.

Specific industrial policies will vary depending on the particular country, with some placing a greater reliance on technologies acquired from abroad through trade and foreign investment, and others exerting greater effort on behalf of local technology development. The balance between the two types of policies may well shift over time as a country familiarizes itself with imported technologies and acquires the capability to replicate, adapt and improve them.

For some developing countries with strong technological capabilities, there may be even scope for pushing the technological frontier outwards. Thus far, there are relatively few examples of developing countries that have established and maintained a strong lead in technologies of global significance, in markets of global scope. This is changing, however, as a number of middle-income developing countries acquire stronger technological capabilities and establish innovation systems.

Some policy steps towards a low-emissions future

In developing countries, there is a need for policies that foster strategic deployment of new technologies, in view of the advantages to be gained by building up new industries and accelerating movement down the learning (and hence, cost) curves. Strategic deployment generally requires a range of incentives, regulation and direct public investment.

Some of the major technologies involved, which are now or could soon be ready for large-scale deployment, include:

- Advanced technologies (such as gasification) for generating electricity from coal and biomass: a suite of technologies whose accelerated deployment will bring higher efficiency, reduced emissions and compatibility with carbon dioxide capture and storage technologies;
- Low-energy building technologies, for markets that are too often impeded by barriers associated with the construction industry and rental markets;
- Primary renewables, notably solar photovoltaics, for which potential scale economies remain large, and wind energy, a proven contributor to emission reductions which involves local learning and related industrial innovation.

A range of government subsidies to producers or users of new technology can be designed to speed technology deployment. Subsidies can take a variety of forms, such as:

- Investment tax credits to firms that bring a new technology to market can lower the upfront investment costs of producing a new type of equipment, and can be tied either to costs or to the level of production. These policies work to increase the supply of a new technology on the market.

- Production tax credits are subsidies granted for a particular type of electricity generation on a per-unit-of-production basis, making renewables such as wind more competitive with respect to higher-emissions production methods.
- To increase demand for a new technology, tax credits or rebates can be granted to purchasers as well as producers, reducing the cost differences between old and new technologies and making the lower-emitting or more efficient new products relatively more attractive. For example, many governments offer tax rebates to consumers who purchase high-efficiency appliances.
- Loan guarantees can subsidize industry by shifting the risk of failure or default to the government, thereby lowering the costs of capital for private firms below what would be available on the open market for an unproved but promising technology.
- Limiting legal liability to the users of a new technology, another implicit subsidy, insulates parties from possible economic damages. This approach may be relevant for carbon capture and sequestration technology, where a release of geologically sequestered CO_2 could potentially undo climate benefits and cause additional harm, giving rise to litigation against the developer.

Energy efficiency

As previously noted, there is a potential for substantial emissions reductions from improving energy efficiency. The building sector, transport and industry appear to offer sizeable opportunities for low-cost improvements; there are also potential, if less well-researched, gains to be reaped in agriculture. There are also other potential benefits from creating jobs in new business activities.

In South Africa, for example, the leading opportunities for energy efficiency include improved building design and improved heating, ventilating and air conditioning (HVAC) efficiency. A "cleaner and more efficient residential energy scenario" involves energy-efficient housing shells, efficiency measures such as deployment of compact fluorescent lamps (CFLs) and geyser (hot-water heater) insulation blankets, and a number of fuel-switching options, including installation of solar water heating, replacement of other fuels with liquefied petroleum gas (LPG) for cooking, and replacement of paraffin (kerosene) with electricity for lighting, linked to substantial increases in residential electrification. Despite the promotion of win-win gains, however, widespread implementation requires some initial investments and efforts to overcome informational, institutional, social, financial and technical barriers.[8]

There are a range of incentives that can reduce initial costs associated with increasing energy efficiency, including subsidies or grants for energy efficiency investments, tax relief for purchase of energy-efficient equipment, subsidies for energy audits, and loans or guarantee funds for energy efficiency projects. Tax incentives, guarantees and other financing measures can help investors overcome the possible constraints on paying the upfront cost of efficiency improvements.

Cleaner coal

Coal is an abundant, low-cost energy resource, but it is also carbon-intensive and polluting. In 2008, the International Energy Agency (IEA) reports, 43 per cent of CO_2 emissions from fuel combustion were produced from coal, compared with 37 per cent from oil and 20 per cent from gas. Under current policies, the IEA has projected, demand for coal will rise by nearly 60 per cent from 2008 to 2035.[9]

Globally, two market problems currently limit the development and adoption of cleaner coal technologies: it costs less to pollute than to control pollution, and barriers such as high development costs slow technological change. Accelerating deployment will require changes at the national and international levels. Businesses want investment certainty through stable policies which recognize the costs and risks of long-term capital investment in pollution control, new combustion designs such as integrated gasification combined cycle plants, and carbon capture and sequestration (CCS) technologies.

Deployment of clean coal technologies must encompass the entire coal supply chain, and parallel progress is needed in technical and non-technical areas for coal to remain an acceptable component in a country's energy mix. A modern coal-fired power plant cannot be considered in isolation from the coal mines, transport infrastructure, and coal markets that supply it. This again underscores the importance of integrated policy responses.

One major challenge will be to develop and deploy CCS systems, a critical technology for coal's long-term future, but one that has not yet been demonstrated on a commercial scale at any coal-fired power plant. This may be an opportunity for some developing countries, and China is already participating in R&D initiatives that aim at accelerating progress.

More broadly, China has an unprecedented opportunity to become a major player in the global market for cleaner, more efficient coal technologies. It has already developed some unique technologies that other countries may want to adopt, and it will certainly create more. It should work with other governments to create a global market for clean energy technologies, and allow its manufacturers to respond with commercially relevant products, for local markets and for export.

Greater efforts in R&D are needed globally, and additional spending alone is not an adequate response to the challenges faced by the energy industry as a whole. China has shown a willingness to participate in international partnerships and joint ventures in many fields aimed at researching, developing and demonstrating new technologies. In the case of cleaner coal, such active participation can speed progress towards those technologies that are most appropriate for commercial markets within China and elsewhere.

Renewables

Strategic deployment of new technologies brings advantages by building up new industries, gaining economies of scale, and accelerating movement down learning curves. At the same time, strategic deployment generally requires regulation, which fosters adoption of technologies that would otherwise be uneconomical; in this way, the benefits of learning by doing and other scale economies are secured.

Even with the expansion of coal consumption in China and India, its growth rate does not match that of renewables, which is doubling every two to five years. In 2010, China became the world leader in total installed wind capacity, adding 18.9 gigawatts of capacity in that year alone, for a 73 per cent increase, and accounting for more than half the world market for new wind turbines. India is also a leader in wind power, ranked No. 5 in 2010, with 13 gigawatts of capacity.[10]

The optimal incentives for renewable energy deployment depend on the technology. The market for solar energy products such as photovoltaic panels, solar water heating, and solar power concentrators ranges from industrial power generation to smaller commercial-scale and domestic installations. Wind power, on the other hand, is almost entirely produced at industrial scale by large companies. Because wind farms are financed by large corporations with access to financial markets, the wind industry has preferred the long-term payback of production tax credits, which provide a return on every kilowatt produced, as a means of making their power more competitive on the market. The biggest concern for smaller-scale solar installations, however, is not the long-term return on the power generated, but rather the initial high cost of installing a system. In this case, an investment tax credit is a better instrument for the industry, allowing producers of solar products to charge lower prices. To make a subsidy cost-effective, care should be given to eliminating free-riders (those companies that would have upgraded their equipment even without a subsidy) and reducing transaction costs.

Other policies that have been used to promote renewables include:

- Feed-in tariffs, as adopted particularly in continental Europe but also in parts of North America and China (see chap. II), which mandate a specific (premium) price to be paid for electricity generated from renewable sources such as wind and solar energy.
- Renewable obligations, known in North America as renewable portfolio standards, which require utilities to source a certain percentage of their electricity from renewable sources.[11]
- Other technology or fuel mandates, such as Brazil's long-standing requirement that cars run entirely or partly on ethanol (see box IV.3), a requirement that has also been established in China.

China now ranks among the top countries in the number of patents for renewable energy technologies. The government had to implement diverse policies to overcome multiple barriers, including the high cost of developing renewable energy, the difficulty of connecting renewable energy to the grid,[12] institutional impediments, a lack of international investment, a weak legal and regulatory framework, and an uncertain level of future demand and thus of prices for renewable energy.

Conclusion

Most developing countries are reluctant to accept binding emissions targets. Their concerns are rooted in fundamental development challenges and reflected in the United Nations Framework Convention on Climate Change, which recognized that countries have "common but differentiated responsibilities". While developed countries are to "take the lead in combating climate change", the UNFCCC says, for developing countries, "economic and social development and poverty eradication are the first and overriding priorities". Developing countries argue that developed countries have yet to demonstrate their leadership in tackling the climate challenge, and that being held to specific emission levels regardless of the economic consequences would be tantamount to putting a cap on their growth and perpetuating unacceptable levels of poverty and inequality.

Establishing low-emissions, high-growth development pathways will be crucial to meeting the climate challenge, reducing global inequality and tackling extreme poverty. If history is any guide, it is unlikely that market forces, by themselves, would be able to establish such pathways and serve as guides through the transition. This chapter has argued that developing countries require the presence of strong and dynamic developmental states capable of providing a coherent vision of the future, of managing the conflicts that arise from change, and of establishing the kind of integrated strategy that will be needed.

Developmental states have managed successful transitions in the past by mobilizing resources and providing missing inputs for productive activities, socializing investment risk, removing barriers, and providing temporary support to those adversely impacted by the shifts in activities. This has involved a blend of pro-investment macroeconomic and industrial policies. Fiscal and monetary policies have given priority to increasing public spending, including investments in energy, education, health and infrastructure. Subsidized credits, credit guarantees, tax breaks, accelerated depreciation allowances, and other policy measures have been used to boost profits in private firms in targeted sectors.

All these elements will certainly be needed to successfully shift to a low-emissions, high-growth pathway. Strategies will have to include a clear vision for energy production and for the energy-intensity of the production structure, for urban development and transportation, and for natural resource use and natural resource intensity of production.

Success will also require collaborations between the state and the private sector. Initial steps can be taken by fostering energy efficiency, implementing cleaner coal processes and developing renewable energy sources. Yet mitigation efforts, no matter how necessary, will not be sufficient to protect developing countries from the threats posed by climate change. The best defence against such threats remains economic diversification to eliminate dependence on a small number of activities that are particularly vulnerable to climatic shocks and changes.

Notes

1. Commission on Growth and Development, 2008. *The Growth Report: Strategies for Sustained Growth and Inclusive Development.* The World Bank, Washington, D.C. http://www.growthcommission.org/index.php?option=com_content&task=view&id=96&Itemid=169.
2. United Nations Conference on Trade and Development, 2007. *The Least Developed Countries Report, 2007: Knowledge, Technological Learning and Innovation for Development.* UNCTAD/LDC/2007. Geneva. http://www.unctad.org/templates/WebFlyer.asp?intItemID=4314&lang=1.
3. World Resources Institute, 2010. 'Climate Analysis Indicators Tool.' *CAIT 8.0.* http://cait.wri.org/.
4. Enkvist, P.-A., Nauclér, T. and Rosander, J., 2007. 'A cost curve for greenhouse gas reduction.' *The McKinsey Quarterly*, No. 1. 35–45.
5. For a more extensive analysis of technological options for sustainable agriculture, see, for instance, chapter III of United Nations, 2011. *World Economic and Social Survey 2011: The Great Green Technological Revolution*, New York. http://www.un.org/en/development/desa/policy/wess/wess_current/2011wess.pdf
6. For an in-depth review of this issue, see Fieldman, G., 2011. 'Neoliberalism, the production of vulnerability and the hobbled state: Systemic barriers to climate adaptation.' *Climate and Development*, 2(2). 159–74. doi:10.1080/17565529.2011.582278.
7. For a more elaborate discussion, see chapter II in United Nations, 2010. *World Economic and Social Survey 2010: Retooling Global Development*, New York. http://www.un.org/en/development/desa/policy/wess/wess_archive/2010wess.pdf
8. Winkler, H. (ed.), 2006. *Energy Policies for Sustainable Development in South Africa: Options for the Future.* Energy Research Centre, University of Cape Town, Rondebosch, South Africa. http://www.erc.uct.ac.za/Research/publications/06Winkler-Energy%20policies%20for%20SD.pdf
9. International Energy Agency, 2010. *CO_2 Emissions from Fuel Combustion 2010 – Highlights.* Paris. http://www.iea.org/co2highlights/co2highlights.pdf
10. World Wind Energy Association, 2011. *World Wind Energy Report 2010.* Bonn. http://www.wwindea.org/home/images/stories/pdfs/worldwindenergyreport2010_s.pdf
11. See box II.1 (pp. 32–33) in United Nations, 2008. *World Economic and Social Survey 2008: Promoting Development, Saving the Planet*, New York. http://www.un.org/en/development/desa/policy/wess/wess_archive/2009wess.pdf
12. This remains a problem with wind power; just under 70 per cent of China's wind capacity was connected to the grid in 2010, according to the World Wind Energy Association.

Chapter V
Technology transfer for climate protection

Introduction

In previous chapters, we have argued for a big investment push to transform energy production and use, and to diversify economies to make them less vulnerable to climatic shocks. That push is to be spearheaded by public investments, but it will be sustained by drawing private investors into an expanding green economy. It must also be accompanied by the technological advances needed to meet mitigation and adaptation challenges. Those advances will entail diffusing existing low-emissions technologies, scaling up new, commercially ready innovations, and advancing new breakthroughs that expand the technological frontier.

A rapid pace of capital formation is often accompanied by accelerated technological upgrading and learning. However, noting the familiar market failures which tend to slow or halt technological progress, chapter IV suggested that more than rapid growth will be needed: a strong public policy agenda mixing price incentives, regulation and interventionist measures, particularly within industrial policy, will be required to ensure a continuous process of technological learning and upgrading. It also suggested that a developmental state will be needed to promote such an agenda in most developing countries. When the required technologies are not available domestically but have to be imported and adapted to local conditions, that agenda becomes more complicated, in large part because the balance between owners and users of technology is tilted in favour of the former.

Technology flows through several well-known channels, the most important being trade, foreign direct investment (FDI) and cross-border technology licensing. Scientific and technical knowledge also flows internationally through research publications, research collaboration and the movement of skilled personnel. Acceleration of the flows of climate-friendly technology raises many of the same issues and challenges facing any sort of technology. What differentiates those technologies from many others is the urgency and scale of the transfers needed to meet the climate challenge. But there is also an underlying ethical challenge posed by climate-friendly technologies, given that the countries most responsible for climate change, or at least their corporations, are set to profit through the transfer of technologies to countries that bear little or no responsibility for the problem.

This chapter is concerned with the international transfer and diffusion of technologies for climate change mitigation and adaptation. The focus is primarily on the "North-South" transfer of technologies, which would allow developing countries to undertake cost-effective actions consistent with—and ideally, capable of reinforcing—their wider economic and social development. It identifies some of the main barriers obstructing such transfer and diffusion and proposes measures for overcoming them. Given the scale and urgency of the climate challenge, it suggests paying more attention to the institutional architecture needed to speed the transition to low-emissions development pathways.

Technology transfer: a global policy challenge

Technology transfer will be fundamental to effective implementation of the United Nations Framework Convention on Climate Change (UNFCCC) beyond 2012. Indeed, there is longstanding international agreement about its essential contribution. As early as 1972, the United Nations Conference on the Human Environment included explicit language emphasizing the importance of technology transfer for the achievement of environmental and developmental goals. The United Nations Conference on Environment and Development, held in Rio de Janeiro in 1992, gave a new urgency to the transfer of environmentally sound technologies for climate change mitigation. Subsequent developments have included agreement on a technology transfer framework at the seventh Conference of the Parties to the UNFCCC, in 2001, covering these key themes and areas for action:

- ***Technology needs and needs assessment***: a set of country-driven activities to identify mitigation and adaptation technology priorities, particularly in developing countries.
- ***Technology information***: the means, including hardware, software and networking, to facilitate the flow of information between different stakeholders to enhance the development and transfer of environmentally sound technologies.
- ***Enabling environments***: government actions which are essential to promoting public and private sector technology transfer, such as fair trade policies, removal of technical, legal and administrative barriers, sound economic policy, regulatory frameworks and transparency.

- ***Capacity-building***: building and enhancing existing scientific and technical skills, capabilities and institutions, particularly in developing countries, to enable them to access, adapt, manage and develop environmentally sound technologies.
- ***Mechanisms for technology transfer***: support for financial and institutional activities to enhance the coordination of the full range of stakeholders in different countries, to engage them in cooperative efforts through technology cooperation and partnerships, and to facilitate the development of projects and programmes to support such ends.

Later meetings and agreements, including the 2007 Bali Action Plan and the 2010 UNFCCC agreement reached in Cancun to establish a multilateral technology transfer mechanism, have further elaborated the agenda for technology development and transfer to support mitigation and adaptation.

The discussion on promoting technology transfer to tackle the climate challenge has evolved quite separately from discussions of technology transfer to meet development goals. The former has focused on quickly putting technologies to widespread use, in developed or developing countries, through learning and adaptation, and addressing market failures that could get in the way. The development discussion, meanwhile, has recently focused on protecting the international position of the creators and owners of technology by linking intellectual property rights to multilateral trade rules and bilateral negotiations. In this context, the degree of protection of the owners of knowledge is often taken as a measure of countries' commitment to good governance and an indication of the attractiveness of their investment climate to foreign firms, whose presence is seen as the surest guarantor of technology access.

However, neither perspective fully acknowledges the urgency of the technological challenge or links it to a big push onto a new low-emissions growth path. Building momentum towards a low-emissions future will require drawing on a variety of mechanisms at the international level, as well as determined leadership that puts collective security before narrow commercial interest.

Intellectual property rights

Incentives or obstacles

Intellectual property rights raise the costs of accessing technology; will this become an important barrier to technology transfer? The answer will depend, among other things, on whether the particular technology has cost-effective substitutes, and on the degree of competition in the industry, which can affect the price of and the terms for licensing. Moreover, the technology covered by an individual patent may provide only a partial capability for exploiting an innovation; total capability may depend on technologies protected by multiple patents or a combination of patented technologies and other forms of knowledge—such as trade secrets and firm-specific know-how, including knowledge embodied in skilled personnel.

There is vigorous debate over whether intellectual property rights, on balance, help or hinder technology transfer. The evidence is inconclusive, and results vary by industry, depending on characteristics such as market dynamism, technological sophistication, importance of RD&D, and ease of imitation and market entry. There is also variation according to level of economic development. In high-income countries, stronger patent rights have been associated with higher levels of productivity, RD&D expenditures, trade flows, FDI and sophistication of the technologies transferred. However, even among these countries, it is unclear whether intellectual property rights are a cause or an effect of these outcomes. Similarly, weak intellectual property rights in the least developed countries tend to be associated with low levels of RD&D and FDI inflows; cause and effect are again difficult to distinguish. Even when technology is transferred to the least developed countries, the principal constraint on its wider use tends to be limited absorption capacity.[1]

Given that stronger protection of intellectual property rights raises the costs of obtaining technologies, it has generally been accepted that low-income developing countries should be exempt from strong intellectual property rights-related obligations, and that the strength of those obligations should only rise with levels of development. Yet, because the current international regime is weighted towards owners rather than users of technology, a graduated approach is likely to support large-scale technology transfer only if it is accompanied by complementary measures with respect to financing, RD&D and technical cooperation—which has not been the case in recent years.

The potential trade-off between intellectual property right protection and technology development and transfer is crucial in the context of climate change. As is clear from figure V.1, the distribution of patent ownership of climate-related technologies is heavily skewed in favour of advanced economies. However, a 2007 analysis found mixed evidence of the importance of intellectual property rights in technology transfer.[2] Based

Figure V.1
Share of patent ownership in the areas of renewable energy and motor vehicles abatement among selected countries, 2000-2004

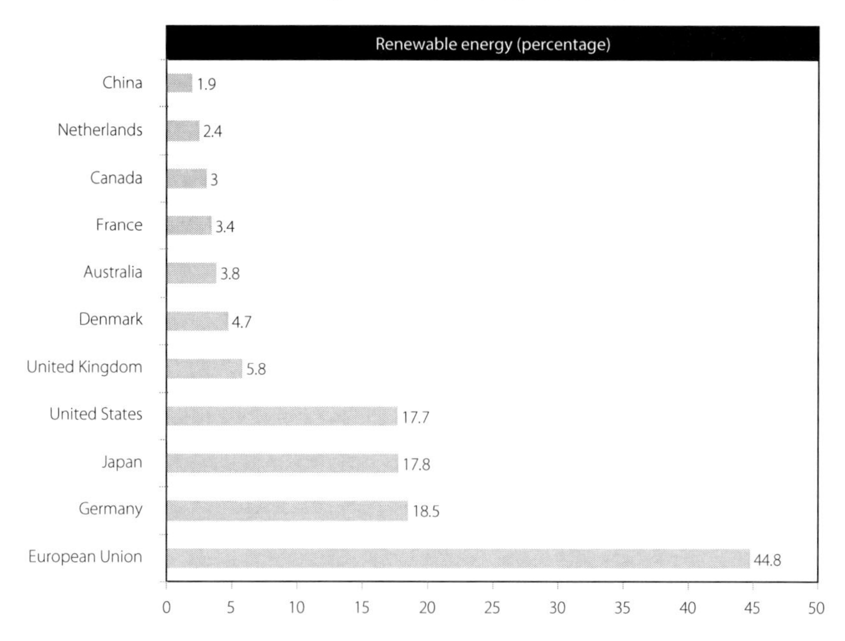

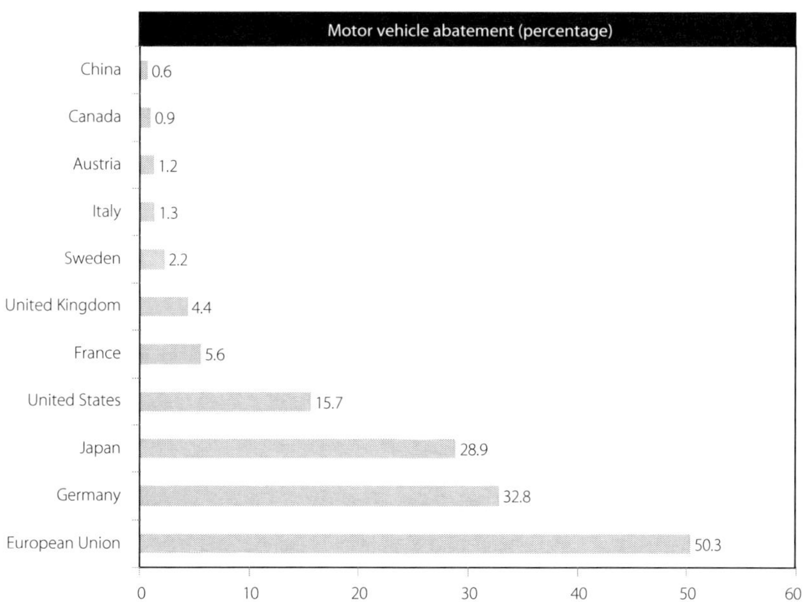

Source: Organization for Economic Cooperation and Development, 2007.

on the examination of three sectors (photovoltaics, wind and biofuels), it concluded that, rather than basic technologies, what are usually patented are specific improvements or features. What matters more are other market distortions. Thus in the photovoltaics sector, for example, where there are multiple small players and no clear leader, developing nations such as India and China have been able to enter and compete in the market. Intellectual property rights also do not appear to be barring developing countries from accessing current-generation biofuel technologies, as shown by Brazil, Malaysia, South Africa and Thailand, among others. More significant barriers and distortions are likely to be associated with industries in which a small number of transnational corporations hold substantial market power.

The most significant barriers and distortions are likely to be associated with the market power of a small number of producers, as in the wind sector. Existing industry leaders are hesitant to share cutting-edge technology out of fear of creating new competitors. China and India, which have significant bargaining advantages, have succeeded in building important firms over the past decade. Whether other developing countries will be able to replicate that success is uncertain.

A much harder question is what lies ahead. To the extent that developing countries make a big investment push to establish a low-emissions development pathway, the market for new technologies can be expected to expand rapidly. However, unanticipated obstacles—particularly the emergence of new sectors linked to these technologies—could slow that transition, or necessitate large shifts of resources to already advanced economies through technology payments.

Multilateral actions to accelerate technology transfer among countries can be of several sorts: those that exploit existing flexibilities of the Agreement on Trade-related Aspects of Intellectual Property Rights (TRIPS), those that require a modification of that Agreement and other disciplines in the framework of the World Trade Organization (WTO), and those that are not necessarily linked to the multilateral trade framework, including initiatives to foster technology-related absorptive capacity and innovation in developing countries through international cooperation.

The flexibility of the TRIPS Agreement

The TRIPS agreement is often viewed as favouring the owners of intellectual property, and constraining potential users in developing countries. It is, however, a complex agreement with several flexible provisions that could be exploited to promote technology transfer, including the options of limiting patentability and making use of compulsory licensing provisions.

Limiting patentability

Patentability refers to the boundaries that delineate which inventions can be patented. Article 27 of the TRIPS Agreement states that "patents shall be available for any inventions … in all fields of technology, provided that they are new, involve an inventive step and are capable of industrial application". These relatively loose criteria leave some space for an individual country to formulate its own policy, including limits on patentability that might promote technology transfer and innovation by reducing possible conflict with existing patents.

Based on the stated goals and guiding principles of the TRIPS Agreement, certain technologies could be excluded from patentability, especially those that are deemed necessary to tackle climate change and/or are subject to anti-competitive measures. Examples of such exclusion already exist within the Convention on Biological Diversity and the International Treaty on Plant Genetic Resources for Food and Agriculture. As the ongoing negotiations within the World Intellectual Property Organization (WIPO) of a substantive patent law treaty could eliminate this opportunity, its impact on climate-related technology transfer should be carefully examined before those negotiations are completed.

There are a number of variants on this approach, including worldwide exemption of climate-friendly technologies and products from patenting; exemption from patenting in developing countries only; allowing developing countries to exclude patents for climate-friendly technologies and products, if they so choose; granting of voluntary licences on request, free of royalty; and granting of voluntary licences automatically, with compensation given to the owner of the technology (the latter two being exceptions to patents, not limits to patentability).

These options could be applied on a graduated basis to countries at different levels of development, with more sweeping exemptions for low-income developing countries and voluntary licences for middle- and high-income developing countries. The size of the country could be another criterion for choosing the appropriate option. For a small developing country, regardless of income level, acquiring a licence for climate-related technology may not be profitable unless it is able to use the licence to tap export markets. In that case, the royalty could be reduced or eliminated and/or the exemption from patent rights could be extended to a whole region.

Compulsory licensing

Even when a technology has been patented, articles 30 and 31 of the TRIPS Agreement offer opportunities for allowing use of a patented technology without the consent of the patent holder through compulsory licensing. For example, Article 30 allows unauthorized use of a patented technology by a country when "necessary for the protection of its essential security interests" or "the maintenance of international peace and security"; depending on one's views of the climate crisis, mitigation and adaptation technologies might qualify under this provision.

Article 31 of the Agreement offers another route to compulsory licensing, providing that a country meets two major criteria. First, reasonable efforts must be made to gain appropriate authorization from the patent holder. This requirement may be waived when the country determines (using its own judgement) that a "national emergency" or "other circumstances of extreme urgency" demand unauthorized use without delay. Discussions leading to the recognition of public health exceptions showed some flexibility in interpreting what qualifies as urgent circumstances, opening the door to potential use of these exceptions in the climate change context.

Second, sales of protected assets must be predominantly in the domestic market for the entity granted the exception. Limiting the technology to the domestic market of one (small or poor) country, however, might prevent the capture of economies of scale which would make the technology cost-effective. Recognizing this fact in the public health realm, the WTO in 2003 approved a temporary waiver of the domestic market requirement for pharmaceuticals in countries with insufficient domestic production.

An amendment to the TRIPS Agreement approved by the General Council of the WTO in 2005 would make this waiver permanent, but years later, it has yet to be ratified by two-thirds of the membership. Such a waiver could conceivably be extended to climate-friendly technologies, but it would certainly meet with strong resistance from owners of technologies in industrialized countries, who could lose potential rents. However, it can be argued that such technologies are not currently reaching developing countries, so the loss of rent would be limited.[3]

Modifying the TRIPS Agreement

A new "Declaration on TRIPS and climate change" could clarify existing options under the Agreement and offer new incentives for transfer of environmentally sound technologies. In particular, exceptions for least developed countries and small island developing States could be pursued, given that in these countries, trade and investment flows appear to be relatively unresponsive to protection of intellectual property rights, while the dangers posed by climate change are particularly acute. Such measures should take into account the uncertain and ever-changing nature of the climate change problem and address adaptation as well as mitigation technologies.

Strong pro-competition provisions would also promote technology transfer. TRIPS Article 40 already allows countries to restrict licensing practices that may have an adverse effect on competition. Developed countries could take the lead here in promoting competition by mandating compulsory licensing for climate-related intellectual property rights held domestically. Pro-competition provisions would, however, meet with strong resistance from intellectual property right holders who exert great influence with several members of the WTO.

New licensing guidelines could provide for fixed, moderate fees for licensing environmentally sound, patented technologies. In cases where the patented technology clearly has environmental benefits, the patent holder would bear the burden of proof in demonstrating why compulsory licensing would be unwarranted. A tiered application fee system for intellectual property rights could waive payments for patent holders who authorize transfer of climate-friendly technologies to developing countries.

If the granting of full licences is an unrealistic option, then temporary licences could be granted along the lines established for plant breeders' exemptions and farmers' privileges under the International Treaty on Plant Genetic Resources for Food and Agriculture. For example, intellectual property right holders could provide developing-country users with technologies for a limited period, with the expectation of receiving payment once the technology was "tropicalized"—i.e., adapted to local requirements.

Mechanisms to evaluate progress on technology transfer should be strengthened. Current evaluation procedures offer neither transparency nor a viable enforcement mechanism. In the absence of formal enforcement, "naming and shaming" would at least provide some measure of accountability.

There are, of course, great political difficulties involved in modifying any WTO agreement. Technology transfer measures can often disadvantage patent holders, who have great political influence in developed countries. Moreover, despite the acknowledgment of development goals, the TRIPS Agreement prioritises equal treatment of nations. However, equal treatment of technologies may not be as crucial, as evidenced by the progress on essential medicines. Global action to address climate change is not a zero-sum game; any country hoping for modification of the TRIPS Agreement in this area will need to stress common interests in advancing the global public good of a stable climate.

Further options for addressing intellectual property rights and innovation incentives

The institutional role of the WTO in the area of climate change is just beginning to be explored. However, the potential clash between trade rules and climate concerns raises serious issues, particularly for developing countries. Other proposals for facilitating technology access and diffusion, which may or may not be consistent with a WTO framework, include:

Open-source information access and increased sharing of public RD&D results

Limited access to information on available technologies is a constraint on technology transfer and adoption. One proposed solution, discussed as far back as 1992, is to establish an information access agreement or an information clearing house of climate-friendly technologies. Some efforts have been made by the UNFCCC on technology transfer and technology needs assessments, but they need to be expanded and better integrated with wider development initiatives.

The Multilateral System of Access and Benefit-Sharing of the International Treaty on Plant Genetic Resources for Food and Agriculture could also be a model for an agreement on access to climate-friendly technologies. Along these lines, a 2006 proposal called for a formal agreement on access to basic science and technology "to ensure widespread access to essential scientific results and to enhance the transfer of basic technological information to the developing world at reasonable cost".[4] As a WTO agreement, this instrument could take advantage of the dispute settlement mechanisms and other institutional structures.

One difficulty with such an agreement would be the challenge of drawing an acceptable line between "basic" and "applied" research. The definition of what is "basic" (and therefore eligible for low-cost transfer) could be construed more broadly in the context of global public goods, so as to favour climate-friendly technologies. In borderline cases, guidelines concerning which research results were confidential and which could be made public would need to be established.

Dedicated funding mechanisms

National governments can subsidize technology development and transfer, through direct subsidies, tax breaks and other fiscal incentives. They can direct the focus of private firms towards particular sectors, such as those encompassing climate change-related technologies, by reducing the risk level of RD&D projects. However, the financial impact of individual governments is limited. Moreover, such expenditures can be exploited by "free riders" on the global level.

A coordinated international funding mechanism would help solve the free-rider problem. Possibilities in this regard include a trust fund encouraging RD&D directly in developing countries, a patent acquisition fund established to buy intellectual property rights from patent holders, and a fund that covers the difference in cost between environmentally sound technologies and business-as-usual technology for developing-country firms. An example of the latter type of mechanism is the Multilateral Fund for the Implementation of the Montreal Protocol, which helped finance the phase-out of ozone-depleting substances.[5]

Another alternative which could circumvent intellectual property rights problems is a technology prize system. Within such a system, the performance characteristics of a desired technology would be defined, a contest would be announced for its development, and a prize would be awarded to the successful innovator in exchange for the intellectual property rights. Prizes help to reduce wasteful spending on marketing and to lower incentives for anti-competitive behaviour; they work best with a specific, clearly defined objective, such as creating a vaccine for a specific disease.

Technology development and transfer mechanisms

An international technology transfer mechanism could be established under the auspices of the Conference of the Parties to the UNFCCC. Supported by a secretariat and expert panels, it could examine the technology challenge in developing countries and, where appropriate, provide technical assistance on the range of technology options available for mitigating and adapting to climate change. This model has been successfully employed within the Montreal Protocol and could be applied to the climate change context.

At regional and national levels, centres dedicated to low-emissions technology innovation and diffusion could be created and linked to and through the international mechanism. They would have an important role to play in making technologies accessible and affordable in developing countries. At least in the initial stages, these centres are likely to be publicly funded, though the precise mix of donor, public and private funding would vary across countries and over time. The appropriate mixture of basic research, field trials, business incubator services, venture capital funding, technical advice and support, and policy and market analysis will depend on local conditions and challenges.

Technology transfer through investment

Accessing clean technologies through foreign direct investment

Many descriptions of foreign direct investment (FDI) emphasize that it is the exploitation of firm-specific advantages, including intellectual property and leading technologies, that allows large corporations to undertake risky and costly activities outside their immediate domestic and regional locations. Hosting foreign firms is seen as a way for developing countries to close the technological gap with more advanced countries. To attract those firms, developing countries have liberalized rules on FDI and trade and, in some cases, created export processing zones, with the expectation that this would help break both technological and foreign-exchange constraints on growth. The results have often been disappointing, particularly in cases where FDI has been a substitute for local domestic capacity-building efforts.

While technology may be physically transferred from the home to the host country through FDI, this does not determine what sort of linkages the transfer creates with the rest of the host economy. How large are the technology spillovers, and do they, as Albert Hirschman asked in 1971, act as "a spur to the expansion of missing local inputs" or "harm the quality of local factors of production"?[6] Answering these questions in greater detail would require a long detour, but two broad findings stand out in the research literature. First, FDI tends to be a lagging variable in the growth process; it is attracted by various factors such as market size, presence of suppliers, human capital, etc., which are the result of a successful development push. That is, FDI is more reliably a follower than a leader of the development process. Second, even when FDI does materialize, active policies are needed to ensure that there are valuable spillovers into the local economy.[7]

Those spillovers can occur in a number of ways: through the movement of skilled personnel between a multinational subsidiary or joint venture and other firms; through technology imitation by competitors; and through technology sharing with suppliers, customers or business partners.

Strong intellectual property right protection is not necessary for extensive foreign investment to occur, as China clearly demonstrates. The country's large market and rapid growth have compelled foreign companies to invest, even at the risk of losing control of proprietary technologies. Countries with "weak" intellectual property right regimes, such as the Republic of Korea, Taiwan Province of China and Brazil in the pre-TRIPS Agreement era have been major technology borrowers, reaping substantial benefits from spillovers from multinational corporations.

Recent research on FDI as a vehicle for technology transfer has pointed to conditions that influence the extent of knowledge spillovers. One study used industry panel data from Indonesia to examine knowledge spillovers between subsidiaries of Japanese multinational corporations and Indonesian firms, and concluded that spillovers were significant only when the Japanese subsidiaries had invested in RD&D themselves.[8] A subsequent study found a significant positive relationship in Indonesia between the training investments of local firms and the extent of knowledge spillovers from foreign ones.[9] All of these findings lead to the conclusion that technology or knowledge transfer through FDI is not automatic, but depends on complementary investments by both foreign and local firms.

There has been little research undertaken to date on the role of FDI spillovers in supporting a transformative low-emissions growth path. Kelly Sims Gallagher's 2006 study of China's automotive industry suggests that hosting FDI is, by itself, no guarantee of technology transfer.[10] China's transportation sector has grown particularly rapidly since the early 1980s, thanks in part to joint ventures with foreign automobile companies producing largely for the growing domestic market. This growth has, in turn, contributed to rapid growth in oil imports. Until 2000, the sector was subject to few regulations and standards on emissions; since then, stricter regulations have been introduced in an effort to force foreign firms to transfer cleaner technologies. However, while these firms have introduced more modern pollution control technologies, they have been reluctant to introduce cutting-edge technology, and the overall impact of their efforts has been dwarfed by the scale effect of rising car ownership. The study concludes that market incentives are, by themselves, unlikely to help China jump to the next stage in terms of cleaner vehicles, such as fuel-cell vehicles, given prohibitive prices and the control exerted over intellectual property by foreign firms. Rather, the government will need to consider a more comprehensive and integrated policy approach, one that seeks to bolster local learning in the automotive sector through support for RD&D and engineering training, and efforts to foster demand for cleaner automobiles in response to higher prices and tighter regulations.

CDM and technology transfer

The Clean Development Mechanism (CDM) was established under the Kyoto Protocol to help developed countries meet their emission targets, by encouraging firms in the private sector to contribute to emission reduction efforts through investments in developing countries. Many of these projects involve multinational corporations from advanced countries. It was expected that such private sector transfers would assist in the transfer of environmentally sound technologies to developing countries.

A few studies have tried to determine to what extent technology transfer is actually occurring through the CDM process. A review by the UNFCCC found that of the 4,984 projects that were in the CDM pipeline as of mid-2010 (including 2,389 that had been registered by the CDM Executive Board), at least 30 per cent—accounting for 48 per cent of estimated emission reductions—involved technology transfer. The rates of technology transfer vary considerably by project type, the review found: only 13 per cent of hydro projects, for example, but 34 per cent of biomass energy and wind projects, 82 per cent of landfill gas projects, and all N_2O projects. Four countries—Germany, the United States, Japan and Denmark—produced 51 per cent of the transferred technology, and 84 per cent of the total came from developed countries. Overall, larger projects were likelier to involve transfer, while only 25 per cent of small-scale projects and 27 per cent of unilateral projects did.[11]

So far, the operation of the CDM has been on too limited a scale, and too heavily concentrated in a few developing countries, to allow it to initiate and sustain the kind of big push towards cleaner technologies recommended in this book. Moves towards the creation of a simplified CDM, including sectoral or technological benchmarks, might make it more effective in raising technological standards in the longer run. However, this is likely to take time.

Trade and climate-related technology transfer

Climate policy discussions have led to a revival of North-South trade and environment debates on how to distinguish between legitimate environmental and health protection measures, as allowed in the WTO, and disguised trade protectionism measures. Despite the establishment of a WTO Committee on Trade and Environment in 1994 to address contentious trade and environment issues, such as how to speed up the transfer of environmentally sound technologies while remaining WTO-compliant, not much progress has been made. Given the absence of prior agreements, more trade disputes are to be expected on measures to account for the carbon-intensity of traded goods and on subsidies to encourage the development of lower-carbon energy sources.

Both the UNFCCC and the WTO recognize that advanced and developing countries are not in equal positions—the UNFFCC through its principle of common and differentiated responsibilities, and the WTO through the principle of special and differentiated treatment for developing countries. In practice, this means that, for instance, under the Kyoto Protocol, developing countries did not have binding greenhouse gas reduction commitments, although they were required to collect data and undertake mitigation and adaptation measures. The level and extent of those measures, however, depended on promised financial, technological and capacity-building support from developed countries.

Along with intellectual property policy changes, some have proposed faster liberalization of trade in climate-related environmental goods and services. In particular, there have been proposals to revisit the Agreements on Subsidies and Countervailing Measures, contained in the 1994 Marrakesh Agreement, to allow subsidies that foster investments in low-emissions technologies.

Trade is important because imported capital goods and services provide an additional channel to access environmental technologies and know-how generated in developed countries. Nonetheless, trade liberalization on its own is not sufficient. Indeed, despite unprecedented market liberalization under the WTO in recent years, and existing technology transfer commitments within the UNFCCC, evidence of technology transfer is slim. Liberalization of trade in environmental goods and services has been slowed by a number of factors, including inaction on the WTO's Doha Development Agenda, the lack of agreement about what constitutes environmental goods and services, and the different views held by the North and the South regarding which tariffs should be lowered more quickly.

Liberalization of trade in climate-related environmental goods and services

In general, developing countries rely much more on tariffs to generate revenues than do developed countries, which have the institutions in place to collect income and sales, or value-added, taxes. The same was true of today's high-income countries at an earlier stage of development: tariffs accounted for more than half of the United States Government revenue in all but 10 of the 122 years from 1789 through 1910.[12] For developing countries that are still in the tariff-dependent stage, significant reduction of tariffs means lower revenues for investment in social and infrastructure development.

With respect to liberalization of environmental goods and services, developing and developed countries have different goals. The former want access to adaptation technologies, while also seeking to protect their nascent environmental goods and services industries so they can eventually compete in world markets. In contrast, developed countries currently have comparative advantages in capital- and technology-intensive environmental goods and services, and propose early liberalization of those goods.

Another obstacle to agreeing on a list of eligible environmental goods and services or climate-related technologies is the lack of specificity in the customs data that tracks traded goods and services. The World Customs Organization Harmonized Commodity Description and Coding System (HS) has been harmonized to only a six-digit level. At that level, customs data are still highly aggregated, with a single category, for example, for "pumps for liquid", including those used in manufacturing wind turbines but also those for other industrial processes. Lifting tariffs on the entire category in order to promote wind power would mean extensive lost revenues and would expose local enterprises, often small and medium-sized, to international competition.

Thus, developing countries fear that the negotiations on environmental goods and services are yet another attempt at prying open their markets. Meanwhile, they observe that developed countries have been slow to meet their obligations regarding the technology transfer, capacity-building and financial assistance required to allow developing countries to acquire climate-related technologies.

Developing countries would have more flexibility if WTO commitments did not require them to lower or eliminate tariffs on all "environmental goods". They would then have the option to develop their own industries and products while maintaining appropriate tariffs, or to lower tariffs on certain environment-related products. This is important because, increasingly, WTO tariffs reductions are bound; in other words, once lowered, they cannot be raised again. Without proper safeguards, the accelerated liberalization of tariffs on environmental goods and services would reduce the policy options available to developing countries for promoting local production along their low-emissions development pathway.

Another definitional issue concerns traditional environmental goods and services such as water treatment, waste collection, and pollution control technologies, versus environmentally preferable products. Proposed lists of environmental goods and services have typically emphasized capital- and technology-intensive pollution control products. Environmentally preferable products, instead of providing an end-of-pipe solution to pollution, reduce pollution during the production process or during the use phase of a product. Well-known examples are organic foods and coffee, and goods which are more energy-efficient in use, such as hybrid cars. The debate over environmentally preferable products in the WTO is at heart a debate about whether (and how) the WTO can distinguish between otherwise similar products based on their processes or production methods.

The most-favoured-nation and national treatment principles now incorporated in the WTO prevent discrimination among "like products" originating from different trading partners, as well as between a country's own and foreign products. Developing countries, fearing that developed nations could use processes or production methods as the basis for non-tariff barriers (by imposing high process-related environmental standards hard to achieve by developing countries), have always taken the position that if the end products have the same physical characteristics, then they are "like products" regardless of how they were produced. However, recent dispute panel findings seem to indicate that, as long as measures to protect the environment do not discriminate between domestic and international producers, or among international producers, they are WTO-compliant under a provision of the General Agreement on Tariffs and Trade (GATT) which allows exceptions to WTO trade rules to protect human, animal or plant life or health.

Embodied carbon

As developed countries adopt climate policies, their energy- and carbon-intensive industries fear having to compete with producers abroad who do not face higher energy prices. Developed-country governments may also fear so-called carbon leakage—the relocation of such industries to non-regulated countries, with associated economic costs and no environmental benefit. A number of developed countries have proposed border adjustments that would "correct" for the differential in carbon emitted in the production of imported goods. If all developed countries join a regime of binding quantitative emission targets, then these measures would be directed largely at developing countries. The intention is to encourage them to become part of a regime of binding targets as well.

Legal experts differ on the details, but many doubt that most border carbon adjustments could be compliant with current WTO rules. Developing countries will eventually have to make significant cuts in their emissions anyway, but for reasons outlined in earlier chapters, they cannot be expected to do so on the same terms or in the same time frame as developed countries, or without financial and technological support. Using stronger measures to try to force developing countries to take on binding commitments is likely only to erode trust between North and South, especially as developed countries have yet to deliver substantial financial and technological support for emission reduction.

Border adjustments are also unlikely to achieve their goal. Only a few energy-intensive sectors (steel, aluminium, paper, chemicals and cement) would be affected, and these are only responsible for a small proportion of economic activity in the developed world. Also, if the border measures cover only basic materials such as aluminium, they hurt the domestic producers that use this input in their processes. If, instead, border adjustments cover manufactured goods such as aluminium-frame bicycles, it would be virtually impossible to calculate the appropriate adjustments on the vast number of products involved in international trade.[13]

Nonetheless, environmental standards can be effective industrial policy instruments for accelerating technological transformation. At present, technical standards are often determined by governments (unilaterally or through agreements among a reduced number of countries) or set by private companies. Wider participation of all parties in the setting of these standards, especially developing countries, should guarantee that environmental standards (including through green labels and ecological footprint certificates) do not become a means of promoting unfair trade protectionism. The Montreal Protocol process, which identified the substances that would be banned and the pace of their elimination, may again serve as an example: the Montreal Protocol provided financial support for adjustment to the agreed standards. A similar, but much more ambitious mechanism to compensate countries economically disadvantaged by the standard-setting of other countries could be devised as part of global financing mechanisms for sustainable development. In addition, a global technology policy mechanism, such as the Technology Executive Committee established by UNFCCC in Cancún (Mexico) in December 2010, could serve as an appellate body where parties adversely affected by such standard-setting could seek a ruling from an expert panel on whether the standards were supported by scientific and practical consideration, as opposed to being protectionist actions.[14]

Low-emissions energy subsidies

The energy sector produces two thirds of the greenhouse gas emissions that cause climate change; along with taxing or capping CO_2 emissions from fossil fuels, many policymakers support subsidies to alternative energy sources. Such subsidies may also raise questions of WTO compliance; however, this issue may be easier to resolve than others because there is a precedent. There was an exception for environmental subsidies under the Agreement on Subsidies and Countervailing Measures; it lapsed in 1999 but could be revived to allow climate-related subsidies that do not injure competitors in other countries.

If the exception for these subsidies could be renewed, both developed and developing countries would be allowed to subsidize general research on climate mitigation and adaptation, without fear of trade sanctions. It is unclear whether free allocation of emissions allowances in carbon trading systems—as proposed in the United States Congress in recent years—would be considered subsidies under the Agreement on Subsidies and Countervailing Measures; a 2009 review found no jurisprudence on this point.[15] It is worth noting that under that Agreement, a country's failure to take action on climate change would not constitute "a subsidy".

The multilateral investment agreement, the Agreement on Trade-related Investment Measures, raises few concerns in this realm, but the 2,500 bilateral investment agreements and the bilateral investment chapters in regional trade agreements contain strong measures—in some cases, perhaps stronger than intended. In the North American Free Trade Agreement (NAFTA), expropriation was so broadly defined and led to so many arbitrations that the United States, Canada and Mexico agreed to clarify and limit the definition of which investors could claim it. These arbitrations have, in some cases, had a chilling effect on countries considering stronger regulations. The fear is that investors could claim that the new regulations constituted unfair and inequitable treatment. Clarification on which climate-related investments could constitute indirect expropriation would give countries the freedom to adopt appropriate regulations without the fear of having to pay excessive compensation to foreign companies.

International measures for capacity building

Technology absorption requires investment in both physical and human capital; the faster the pace of capital formation, the greater the potential rate of technological change. In this process, promoting local technological learning and capacities will be critical. Technology needs will differ from region to region, but in all cases, active government policy will be a component of successful outcomes (see chap. IV). Moreover, the global nature and urgency of the climate challenge imply that the rapid dissemination of appropriate technological options will require international collaboration.

This is particularly true in the area of RD&D, where developing countries lag significantly and risk falling further behind as new technologies emerge. Examples of technologies that will be critical to a new development pathway include carbon capture and sequestration (CCS), low-emissions biofuels, and breakthroughs in renewable energy such as photovoltaic panels. Moreover, developing countries also need access to best practices with respect to adaptation technologies, in the areas of agriculture, disaster management and urban planning. These technologies are often closely interrelated and can also address food and energy security. Developments in all these areas are best addressed through a structured global programme and funding, to develop greater coordination among the myriad research institutions working to meet these challenges, and to ensure the widest dissemination of the results (box V.1). Transparent and readily accessible research is all the more important because regulatory and legal frameworks, such as standard-setting, are likely to emerge on the basis of these results.

Particularly with respect to cutting-edge technologies, well-educated engineers and managers are essential. Enhanced education and training programmes are needed in the areas of technical, administrative, financial, regulatory and legal skills. Along with improving domestic educational programmes, developing countries can guard against a "brain drain" by offering incentives such as wage flexibility, repatriation grants, and support for technology start-ups. Developed countries, for their part, should subsidize offshore training, conference attendance and, in some cases, temporary employment for graduates from developing countries. Grant proposals for research on environmentally sound technologies involving developing-country teams could also receive special consideration. Capacity-building might also be pursued through the cooperation agreements that increasingly accompany regional trade agreements among OECD countries. These would help developing countries conduct an assessment of the obstacles to their low-emissions energy development. Aid-for-trade programmes should also be tapped in this regard.

Box V.1: Intellectual property rights and publicly funded technologies

The issue of publicly owned technology transfer was addressed at the United Nations Conference on Environmental Development, held in Rio de Janeiro in 1992. The agenda included a statement that governments and international organizations should promote the "formulation of policies and programmes for the effective transfer of environmentally sound technologies that are publicly owned or in the public domain." Implementation of this provision has been very weak.

Developed-country governments sponsor a range of research and development (R&D) activities geared towards climate technologies. For example, in 2010, European governments spent $1.1 billion on R&D in renewable energy, slightly more than the private sector; the United States Government spent $1.0 billion, twice as much as corporations. "Green stimulus" packages in those countries as well as in Japan, South Korea and Australia have given a significant boost to renewables R&D.

A 2005 survey of government-sponsored R&D in the United States, Canada, the United Kingdom, the Republic of Korea and other countries that are members of the Organization for Economic Cooperation and Development (OECD) found that it is a common practice for governments to grant ownership of intellectual property rights (patents, copyrights, trademarks, etc.) to the recipient research institutions. In the United States, for example, government-sponsored research usually ends up being patented.

Given the role that governments play as the main driver of R&D for climate technologies, it will be necessary to explore possibilities for the transfer of publicly funded climate technologies to developing countries. In OECD countries, where ownership of most of the technology needed for mitigation and abatement is held, governments are in a strategic position to influence technology flows through their influence on private sector entities or public institutes which receive funding for R&D.

Sources: Sathaye et al., 2005, United Nations Environment Programme and Bloomberg New Energy Finance, 2011.

What is clearly required is a massive international effort. Table V.1 presents various mechanisms to promote technology development and transfer. Three closely related initiatives could plant the seeds of greater international collaboration in this area:

- ***A multilateral technology fund*** to support an international programme on the diffusion of climate technology and to strengthen and coordinate regional and national RD&D efforts in developing countries. Such a fund could be housed in the secretariat of the UNFCCC and draw on the existing network of scholars and scientists within the Intergovernmental Panel on Climate Change (IPCC) in the design of its programmes. It could draw on the experience with the Global Environment Facility (GEF) (see box V.2). A comprehensive programme would need to focus on the full range of technological challenges throughout the basic science, applied research, demonstration, deployment and commercialization stages of

Table V.1
Innovative mechanisms to promote technology development and transfer

Mechanism	*Rationale*	*Issues to consider*
Publicly supported centres for technology development and transfer	The Green Revolution model: makes technologies available to developing countries without intellectual property rights protection	Suitable for mitigation, or only for adaptation technologies?
Technology funding mechanism to enable participation of developing countries in international R&D projects	Resultant intellectual property rights could be shared; patent buyouts could make privately owned technologies available to developing countries	Is there sufficient incentive for participation by developed-country private sector technology leaders?
Patent pools to streamline licensing of inventions needed to exploit a given technology	Developing-country licencees will not have to deal with multiple patent holders	What are the incentives to patent holders? Would government regulation be needed?
Global R&D alliance for research on key adaptation technologies	Model of research on neglected tropical diseases	Is such an approach suited to mitigation technologies?
Global clean technology venture capital fund	Fund located with a multilateral financing institution which will also have the rights to intellectual property	Will new technology ventures be commercially viable if they do not own intellectual property?
Eco-Patent Commons for environmentally sustainable technologies	Private sector initiative to make certain environmentally sound technologies available royalty-free on a "give-one, take-one" model	Voluntary, private incentives appear weak. What about those companies without a patent to contribute?
Blue Skies proposal of European Patent Office: differentiated patent system with licensing of rights to climate change technologies	Complex new technologies based on cumulative innovation processes need to be treated differently from, for example, pharmaceuticals	Appears to address concerns similar to those addressed by the patent pools proposal: more specifics needed on implications for technology access
More favourable tax treatment in developed countries for R&D performed in developing countries	More proactive, technology-push approach by developed-country governments	May face domestic political constraints
Technology prizes	Reward innovation without awarding intellectual property rights to innovators	Requires a well-specified research objective

Source: United Nations Department of Economic and Social Affairs, 2008.

Box V.2: The Global Environment Facility

Technology transfer will play a critical role in the global response to climate change, and promoting it is one of the commitments embodied in the United Nations Framework Convention on Climate Change (UNFCCC). Parties to the Convention agreed to provide funding to support these goals, and the Global Environment Facility (GEF), an independent entity that already served as the financial mechanism for other environmental programmes, was chosen to administer the new funds.

Established in 1991, the GEF is the world's largest funder of environmental projects. As of early 2011, it had allocated $9.2 billion—supplemented by more than $40 billion in co-financing. Climate change projects are among the biggest funding areas, with $2.8 billion invested from 1991 to 2009 (through UNFCCC and other funds). Most financing is in the form of grants to developing countries and countries with economies in transition. Through its Small Grants Programme, which typically awards $20,000 to $35,000 at a time, the GEF has also made more than 13,000 grants, worth more than $304 million, directly to non-governmental and community organizations. Below are some examples of technologies supported by the GEF.

Energy-efficient lighting and appliances

The GEF has built a portfolio promoting energy-efficient appliances and technologies in developing countries. Supported interventions typically focus on instituting energy-efficiency standards and labels, consumer education, and testing and certification of appliances. In countries where there is substantial manufacturing capacity, the GEF has also supported enterprises in developing new energy-efficient appliance models and in acquiring technical information and knowledge from more advanced countries. In Tunisia, for example, a GEF project helped 10 out of 12 local appliance manufacturers develop more energy-efficient models.

Industrial energy-efficiency technologies

The GEF has funded industrial projects in several countries to promote technology upgrading and the adoption and diffusion of energy-efficient technologies. Some projects focus on the development of market mechanisms such as energy service companies, the creation of dedicated financing instruments, and technical assistance to stimulate investments in new technologies. Others are designed to identify one or more subsectors where specific technologies can be promoted. The range of industries includes construction materials (brick, cement and glass), steel, coke-making, foundry, paper, ceramics, textiles, food and beverage, tea, rubber and wood. A number of projects also promote energy-efficient equipment such as boilers, motors and pumps, as well as cogeneration in the industrial sector. In some projects, the GEF has promoted South-South technology transfer; for example, energy-efficient brick kiln technology was brought from China to Bangladesh.

High-efficiency boilers

The China Efficient Industrial Boilers project received a $32.8 million grant from the GEF to upgrade existing boiler models, adopt new high-efficiency boiler models by introducing modern manufacturing techniques and boiler designs, and provide technical assistance and training for boiler producers and consumers. Completed in 2004, the project successfully worked with nine boiler manufacturers and nine boiler auxiliary equipment makers. With GEF support, the manufacturers in China acquired

Box V.2 (cont'd)

advanced efficient boiler technologies, built prototypes, and began commercial production. Through technical assistance, the project also led to the revision and formulation of national and sector standards, while strengthening the technical capacity of China's boiler sector.

Solar water heaters

Solar water heater technology may seem simple, but it is not. High-quality fittings, solar collectors and installation are required for satisfactory operation. Accordingly, inexpensive materials, poor workmanship and shoddy installation have often resulted in non-functional units and abandonment of installations. GEF experience has shown that knowledgeable staff and the observance of high standards are critical to the successful dissemination of this technology. In Morocco, for example, a GEF project supported upgrades to older solar installations, training for technicians to ensure higher-quality new installations, and subsidies that revived the market.

Waste to energy

Several projects that support the use of methane from municipal waste have qualified for GEF funding as both renewable energy projects and short-term response measures, because of their cost-effectiveness. The India biomethanation project, for example, was designed to exploit India's endogenous capacity to adapt and replicate biogas technology for industrial wastes. The GEF played a role in helping increase the uptake of such technologies, which have gone on to become highly profitable when implemented under the Clean Development Mechanism.

Concentrating solar power

The GEF, together with India, Mexico, Morocco and Egypt, developed a portfolio of four concentrating solar power demonstration plants. The projects built solar fields, typically with 30 MW capacity, as part of hybrid gas-turbine plants. Successful hybridization of the gas-turbine and solar power plants would enable the projects to dispatch power at will, thereby making them more economically attractive.

Source: Global Environment Facility, 2011, and other GEF materials.

cleaner technologies. Critical technologies such as carbon capture and sequestration and the next generation of biofuels, in which developing countries have a particular interest, would have to be high up in the agenda. Given the public nature of RD&D, it would be essential to ensure dedicated and predictable financing for such a fund, using the kinds of instruments discussed in chap. VI. Such a fund could act as a focal point for the coordination of ongoing research in climate technologies at the international and national levels and among public, private and non-profit organizations, while ensuring open access to all available research in line with the urgency of the challenge.

- ***A human skills transfer programme.*** A scaled-up human capacity development effort could consist of a temporary (perhaps only a virtual) movement of skilled unemployed/underemployed workers from developed countries (engineers, technicians, primary education teachers, experts in sustainable agriculture, and qualified blue- and white-collar workers) into developing countries to provide technical and vocational training. This could involve "reverse outsourcing", that is, programmes utilizing the Internet and other communications technologies to provide long-distance training services in critical areas for developing countries. During a recession, many highly skilled technicians, teachers and professionals are laid off. Even if only a small fraction of them participated in a technology transfer corps for six months to two years, a significant transfer of skills and know-how could be achieved. This would be a win-win solution for developing countries requiring more help and for cash-constrained developed countries obliged to pay unemployment insurance.

- ***A public technology pool.*** The results of fully funded public research on climate technologies should not be the basis of private patents: they should be made available at low or no cost to all countries. A technical secretariat would be needed to monitor, collect and disseminate such research, to act as a clearinghouse for existing publicly funded technologies, and to actively promote access to those technologies, particularly for developing countries. An important step in this direction was made through the UNFCCC agreement, reached in Cancún in December 2010, to establish the Climate Technology Centre and Network (CTCN), an operational body to facilitate networking among national, regional, sectoral and international agencies dealing with climate-related technologies.[16]

Conclusion

Innovative transfer of both technologies and know-how will be required to meet climate change objectives in the context of both mitigation and adaptation. This chapter has identified possible obstacles to the transfer of technology that could arise internationally with respect to intellectual property rights, corporate behaviour and trading rules. Existing agreements sound restrictive, often strongly favouring owners of intellectual property over potential users. To date, these factors have not proved prohibitive. However, they are likely to take on greater significance if developing countries embark on a big push toward a low-emissions, high-growth development pathway.

Anticipating those obstacles and devising ways around them are urgent tasks for the international community. Several creative solutions have been proposed to make use of existing flexibility, and/or to modify the rules that are now in place. These solutions would require consensus, since they might entail the amendment of WTO rules and the adoption of special climate waivers based on the urgency of the rapidly evolving climate situation. They will also need to pay careful attention to the implications of the WTO principles of non-discrimination and of UNFCCC principles, especially that of common but differentiated responsibilities and capabilities. Since any post-2012 agreement is likely to retain these principles, the challenge will be to ensure the coherence and compatibility of their applications. New mechanisms for international support for technology transfer will be needed in order to ensure the diffusion and application of essential innovations in the countries where they are most needed.

Notes

1 For a closer look at this topic, see Blyde, J. S. and Acea, C., 2003. 'How does intellectual property affect foreign direct investment in Latin America?' *Integration and Trade*, 7(19). 135–52. For more on the absorption capacity issue, see United Nations Conference on Trade and Development, 2007. *The Least Developed Countries Report, 2007: Knowledge, Technological Learning and Innovation for Development*. UNCTAD/LDC/2007. Geneva. http://www.unctad.org/templates/WebFlyer.asp?intItemID=4314&lang=1.

2 Barton, J.H., 2007. *Intellectual Property and Access to Clean Energy Technologies in Developing Countries: An Analysis of Solar Photovoltaic, Biofuel and Wind Technologies*. Trade and Sustainable Energy Series, Issue Paper 2. International Centre for Trade and Sustainable Development, Geneva. http://ictsd.org/i/publications/3354/.

3 For more on this topic, see Hoekman, B. M., Maskus, K. E. and Saggi, K., 2005. 'Transfer of technology to developing countries: Unilateral and multilateral policy options.' *World Development*, 33(10). 1587–1602. doi:16/j.worlddev.2005.05.005.

4 Barton, J.H., and Maskus, K.E., 2006. 'Economic perspectives on a multilateral agreement on open access to basic science and technology.' In *Economic Development and Multilateral Trade Cooperation*, S.J. Evenett and B.M. Hoekman (eds.). World Bank and Palgrave Macmillan, Basingstoke, U.K.

5 For an analysis of lessons from the Montreal Protocol implementation for the climate process, see Andersen, S.O., Sarma, K.M., and Taddonio, K., 2007. *Technology Transfer for the Ozone Layer: Lessons for Climate Change*. Earthscan, London.

6 Hirschman, A.O., 1971. *Bias for Hope: Essays on Development and Latin America*. Yale University Press, New Haven, CT.

7 On the links between FDI and development, see Kozul-Wright, R. and Rayment, P., 2007. *The Resistible Rise of Market Fundamentalism: Rethinking Development Policy in an Unbalanced World*. Zed Books and Third World Network, Penang, Malaysia. Chap. 4.

8 Todo, Y. and Miyamoto, K., 2006. 'Knowledge Spillovers from Foreign Direct Investment and the Role of Local R&D Activities: Evidence from Indonesia.' *Economic Development and Cultural Change*, 55(1). 173–200. doi:10.1086/505729.

9 Miyamoto, K., 2008. 'Human capital formation and foreign direct.' *OECD Journal: General Papers*, 2008(1). 1–40. doi:10.1787/gen_papers-v2008-art4-en.

10 Gallagher, K.S., 2006. 'Limits to leapfrogging in energy technologies? Evidence from the Chinese automobile industry.' *Energy Policy*, 34(4). 383–94. doi:16/j.enpol.2004.06.005.

11 United Nations Framework Convention on Climate Change, 2010. *The Contribution of the Clean Development Mechanism Under the Kyoto Protocol to Technology Transfer*. Bonn. http://cdm.unfccc.int/Reference/Reports/TTreport/TTrep10.pdf

12 U.S. Bureau of the Census, 1975. *Historical Statistics of the United States, Colonial Times to 1970, Bicentennial Edition, Part 2*. U.S. Department of Commerce, Washington, D.C.

13 See Cosbey, A. (ed), 2008. *Trade and Climate Change: Issues in Perspective*. Final Report and Synthesis of Discussions, Trade and Climate Change Seminar, Copenhagen, June 18–20, 2008. International Institute for Sustainable Development. http://www.iisd.org/pdf/2008/cph_trade_climate.pdf

14 See also chap. VI (pp. 182–184) of United Nations, 2011. *World Economic and Social Survey 2011: The Great Technological Transformation*, New York. http://www.un.org/en/development/desa/policy/wess/wess_current/2011wess.pdf

15 Hufbauer, G.C. and Kim, J., 2009. 'Climate Policy Options and the World Trade Organization.' *Economics: The Open-Access, Open-Assessment E-Journal*, 29, 18 June. doi:10.5018/economics-ejournal.ja.2009–29.

16 For further discussion, see chap. VI of United Nations, 2011. *World Economic and Social Survey 2011: The Great Technological Transformation*, New York. http://www.un.org/en/development/desa/policy/wess/wess_current/2011wess.pdf

Chapter VI
Financing the development response to climate change

Introduction

There is no way to escape the need for large-scale investments to meet the climate challenge around the world. Developed countries have begun to make adjustments, but the pace has been slow. In 2008 and 2009, expectations were raised by the inclusion of green investments in stimulus packages in response to the global financial crisis, but austerity measures have since put a halt to that trend. Developing countries, meanwhile, have to ensure their climate policies are consistent with long-standing growth and development objectives. The key to this, we have argued, is to adopt an investment-led and integrated approach. In particular, large-scale investments will need to be front-loaded to make a "big push" toward climate change mitigation and adaptation. Yet those investments will involve significant initial costs and a high degree of uncertainty.

The debate about the economic dimension of climate policy has been dominated by assessments of market-based mechanisms such as cap and trade and carbon taxation. There is little doubt that establishing a realistic price for carbon has to be part of any policy agenda, and private investment will, of course, have a major role in any low-emissions economic future. The question, however, is whether market mechanisms alone can induce the required shifts in production and consumption patterns and mobilize the large-scale investments needed. This seems doubtful. Price mechanisms are an unreliable guide in a case like this, where the required investments are very large, economies of scale and learning curve effects are important, and the returns on investment depend both on climate uncertainties, and on a series of complementary investment efforts and policy initiatives. This is even truer today, against the backdrop of systemic financial market failure and volatile carbon markets that are not compatible with long-term investment planning.

While market mechanisms should play a role in a more comprehensive package of measures, the investment path needed to meet the climate challenge will require heavy reliance on regulation and large-scale public investments in order for the necessary economic transformation to take place.

Public investment, financed by both tax revenues and long-term borrowing, has played a transformative role in shaping development pathways in the past, including in today's most advanced economies. In many cases, external financial support has been critical. Achievement of the transition to a low-emissions, high-growth path in developing countries will also require massive public investment, funded to a large extent through external resources, particularly in the early stages. Along with reducing emissions, the aim of such investments will be to crowd in profitable investment opportunities for the private sector along the new development pathway.

Given the great uncertainties involved, it is not easy to define an appropriate financing framework for climate change. Depending on what target is used for stabilizing greenhouse gas concentrations and other modelling assumptions, estimates of the annual cost of mitigation range up to 2 per cent of world output by 2030—though in all cases, doing nothing would lead to much higher economic losses. Adaptation costs are apparently smaller, but also uncertain, depending on assumptions about anticipated climate impacts. On both fronts, large shares of the required investments must be made in developing countries. Yet while at the global level, the costs seem quite affordable, most models do not consider the challenges of increasing investment to such high levels in developing countries. They also fail to consider the possibility that climate-related investments could trigger a high-growth pathway, allowing countries to meet longstanding development goals.

The key questions are, first, which measures will be most effective in both mobilizing the required resources and steering investments in the desired direction, and second, how should the costs be distributed across nations and population groups? Market-based scenarios, such as one developed by the World Bank, assume a rapidly growing role for funding mechanisms such as cap and trade systems and carbon offsets, complemented by moderate amounts of multilateral, mostly developed-country funding; developing countries' role in financing investment remains small through 2050.[1] The high-growth, low-emissions scenario envisioned in this book starts with much larger upfront investments by developed countries to generate a big push in the desired direction, initially led by public investments; but within a few decades, the rapid rate of growth makes it possible for developing countries to finance most of their own ongoing investment needs. By 2050, the requirements for multilateral and developed-country funding are much lower than in a predominantly market-based scenario. This happy outcome is only possible, however, if the initial funding for the big push is forthcoming.

This chapter begins by assessing the likely scale of resources needed to achieve low-emissions, high-growth pathways, and to make vulnerable countries and communities more resilient with respect to climate change and shocks. It then considers how those resources could be mobilized and, in particular, both the advantages and the limitations of cap-and-trade mechanisms and carbon taxes as financing vehicles in the initial years. A wide mix of financial mechanisms will likely be required, including domestic resources. The chapter concludes with a consideration of the elements of an alternative global investment regime, initially dependent on significant public sector involvement and a prominent role for a multilateral financing mechanism.

Estimating financing requirements

The parties to the UNFCCC agreed that developed countries would provide financial resources to developing countries to meet "agreed full incremental costs" of implementing mitigation and adaptation activities as well as related activities encompassing, among others, climate research, training and management of sinks. These are not voluntary commitments, but treaty obligations. However, estimates of those global costs vary widely, depending on model assumptions and the required emissions target, among other factors (see chap. I). What is certain is that the longer the response to climate change is delayed, the more damaging will be the threats to lives and livelihoods, and the greater will be the resources required to respond to those threats. It is also important to recognize that appropriate responses to the climate challenge will not be uniform across all countries. In particular, there will likely be sharp differences between developed and developing economies, given the higher mitigation and adaptation costs facing the latter.

Mitigation costs

Figure VI.1 presents a range of estimates of mitigation costs. Given the uncertainties and unknowns in these costing exercises, it is not surprising to find the range varying from as little as 0.2 to about 2 per cent of world GDP, or between $180 billion and $1.2 trillion per year (by 2030). The range of estimates depends on methodologies used as well as on whether the

Figure VI.1
Range of estimates of annual additional cost of mitigation strategies, 550 ppm and 450 ppm scenarios, world and developing countries

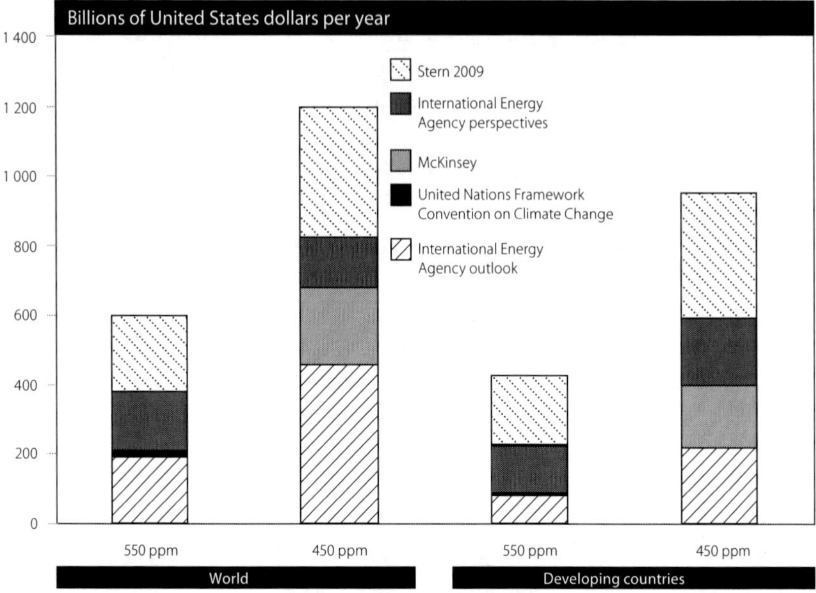

Sources: Stern, 2009; International Energy Agency, 2008a and 2008b; United Nations Framework Convention on Climate Change, 2008; and McKinsey & Company, 2009.

target of stabilization of greenhouse gas concentrations is set at 450 parts per million (ppm) or 550 ppm. In all cases, the costs to the world economy would be considerably higher under a business-as-usual scenario, with increasing expected losses, and a growing threat of an abrupt, irreversible catastrophe, as time goes on.

More than half of the incremental costs of greenhouse gas abatement are expected to fall on developing countries, whose energy investments over the coming decades are projected to grow much faster than those of developed countries (see chap. II). The costs include investments in renewable energy, which at current prices remains more expensive than coal or other fossil-fuel alternatives; more efficient and lower-emitting coal-based power plants, including integrated gasification combined cycles and supercritical coal power plants; carbon capture and storage; and more energy-efficient boilers, furnaces and other industrial equipment. However, from a development perspective, it is difficult to separate these incremental investments from the challenge of meeting growing energy demand in developing countries, as well as interrelated demands on the transportation system and in urban expansion, improved irrigation and water management, and other sectors.

Adaptation costs

Discussion of adaptation costs has focused on the additional investment needed to increase resilience and reduce the impact of anticipated future damages caused by weather events. In practice, adaptation costs may also include disaster relief expenditures when damages actually occur. However, because these costs depend on the uncertain probability and severity of climatic threats, whose local impact is closely linked to other vulnerabilities, it can be hard to determine where traditional development expenditures end and new adaptation expenditures begin. In any case, adaptation measures need to be embedded in broader development strategies, as discussed in chap. III.

A UNFCCC report estimates that annual worldwide adaptation costs will be $49 billion to $171 billion by 2030 (see table VI.1).[2] Its adaptation scenario covers five sectors, with the uncertainty arising from the cost of adapting infrastructure, estimated at between $8 billion and $130 billion. The World Bank's Economics of Adaptation to Climate Change (EACC) project, meanwhile, which included a global analysis and seven country studies, estimated that adapting to 2° C of warming would cost developing countries $70 billion to $100 billion per year between 2010 and 2050.[3]

Table VI.1
Additional investment and financial flows needed for adaptation in 2030, by sector

Sector	*Areas/adaptation measures considered*	*Global cost (billions of 2005 United States dollars)*	*Proportion needed in developing countries (percentage)*
Agriculture, forestry and fisheries	Production and processing, research and development, extension activities	14	50
Water supply	Water supply infrastructure	11	80
Human health	Treating increased cases of diarrhoeal disease, malnutrition and malaria	5	100
Coastal zones	Beach nourishment and dikes	11	45
Infrastructure	New infrastructure	8–130	25
Total		**49–171**	**34–57**

Source: United Nations Framework Convention on Climate Change, 2008, table 5.

The financing challenge

The estimated additional investments needed for adaptation and mitigation to address climate change are large in absolute terms, but only a small fraction of world output (on the order of 1 to 2 per cent)[4] and of estimated total global investment (2.5-5.0 per cent) in 2030. There is a growing recognition, however, that many of these investments need to be front-loaded, in order to accelerate the shift to a low-emissions economy and to minimize the damage from unavoidable changes in the climate. Front-loading implies more pressure on the financial system to quickly mobilize resources. Moreover, as discussed in earlier chapters, these investments in adaptation and mitigation are often closely interrelated and will make sense only in combination with complementary investments designed to meet wider development objectives, such as developing infrastructure, raising agricultural productivity and diversifying economic activity.

Despite the recent proliferation of climate-related funds, the amount currently promised and expected to be available for meeting the climate challenge in the near term is woefully inadequate. A United Nations Environment Programme report estimates 2009 climate finance flows at about $16 billion, including $13 billion from four bilateral institutions.[5] Under the Copenhagen Accord, developed countries agreed to jointly mobilize $100 billion a year by 2020 “from a wide variety of sources” to address the climate mitigation and adaptation needs of developing countries. They also agreed to make $30 billion a year in “Fast-Start” funding available between 2010 and 2012—though questions have arisen about whether these will truly be new funds or replace existing aid, and whether economic problems would lead some countries to scale back their pledges. Even in the most optimistic view, however, the funds available for adaptation are expected to fall far short of projected needs.[6]

This suggests that global financing for climate change will require a much more determined effort on the part of advanced countries to provide bold leadership on climate and bolster international cooperation. But it will also require an effort on the part of developing countries to mobilize a larger share of their resources for cleaner investments along a new, sustainable growth path.

The purpose of a substantial injection of external financing is to not only give a big push to low-emissions development, but also to accelerate and sustain growth in developing countries. As discussed in earlier chapters, those external investments, combined with domestic policies such as price incentives, regulation and targeted industrial policies, would in turn spur investment from domestic sources in both the public and the private sectors. The evolving mix of public and private investment will vary, but for many developing countries, and possibly for some developed countries, public investment will have to take the lead, along with stronger regulations, before large-scale private investment begins to materialize.

Crowding in private sector resources

A clear objective for climate change policy is to reveal the hidden costs of high-emissions technology by putting a price on carbon. In the case of adaptation, incentives will likely involve the sharing of costs among consumers, the private sector and governments. Insurance markets offer one way to share costs, and other innovative instruments have been introduced in recent years. However, these instruments still operate on a very limited scale, even in more advanced countries, and tend to be particularly expensive in developing countries, where coverage is very limited.

Some consumers are adjusting their consumption patterns to reduce their carbon footprints, and some companies are voluntarily reducing their emissions as well—though given the extra cost, it is not surprising that they remain in the minority. Without more aggressive government intervention, it is unlikely that these trends will make a significant impact on global emissions.

This section reviews a range of mechanisms that fall broadly in the category of market-based measures, as their main focus is on changing the price of carbon to reallocate resources away from emission-intensive forms of energy. Several of these mechanisms are also expected to mobilize resources for financing investments in greater energy efficiency and use of renewable energy.

Market-based incentives in developing countries

The economic policy debate on climate change has been dominated by the search for market-based solutions to acknowledged market failures. The idea is to internalize climate-related costs that are now considered externalities, so that greenhouse gas emitters recognize and pay for their emissions and have an economic incentive to reduce them. If they are efficient, it is assumed that they will seize all investment opportunities for cutting emissions that cost less than the established price for carbon.

In addition to its intended purpose, carbon pricing has important, generally undesirable side effects. It affects the level and distribution of real household income, both directly through a household's use of fossil fuels and indirectly through the prices of other commodities; several studies have found that low-income groups are disproportionately affected. One way to address this is to introduce differentiated pricing by, for example, increasing prices commensurate with the amount of energy used; alternatively, compensatory mechanisms in the form of subsidies for lower-income groups could be put in place. In the United States, some have proposed returning a large share of carbon revenue to households as a per-capita dividend.

The principal categories of market-based incentives are cap and trade systems, carbon taxes, pricing ecosystem services (as in schemes to reduce deforestation), and other incentives for targeted investments.

Cap and trade

Pricing greenhouse gas emissions as a pillar of mitigation policy emerged in the early 1990s with the UNFCCC and took on greater prominence with the Kyoto Protocol. The Protocol, adopted by the Conference of the Parties to the UNFCCC in 1997, set differentiated, legally binding targets for industrialized countries, while setting up an emissions trading scheme to meet those targets. A financing mechanism for projects in developing countries, the Clean Development Mechanism (CDM), was launched at the same time.

These mechanisms are designed to work with a cap-and-trade programme, where governments set an overall emissions cap and then issue tradable permits (either by auction, or to selected firms or organizations) which allow their holders to emit a specified quantity of greenhouse gases. Firms that can reduce their emissions more cheaply can sell their allowances, or refrain from bidding for them at auction, creating a competitive market which reveals the market-clearing price for allowances. Yet the global carbon market remains small, valued at $142-144 billion in 2009-2010, driven almost entirely by the European Union's Emissions Trading Scheme (EU ETS).[7] Both the price of carbon and the volume of trading have been unstable: an assessment of the European Union (EU) experience with emissions trading between September 2005 and March 2008 found that the price of carbon was more volatile than stock market indices—a problem that could discourage long-term investment.[8]

Although trading to date has been largely confined to the EU, developing countries can be pulled in indirectly through CDM funding for emissions-reducing projects. As of June 2010, the CDM Executive Board had registered 2,389 projects, and another 2,595 were under review.[9] But though there are now CDM projects in about 70 countries, four countries host the vast majority: China, 41 per cent; India, 22 per cent; Brazil, 7 per cent; and Mexico, 5 per cent.[10] However, problems with the programme, long delays in the approval of credits, and questions about the CDM's post-2012 future, along with the global financial slump, have weakened the market; the value of the primary CDM market peaked in 2007, at $7.4 billion, and fell to $1.5 billion in 2010, well below the 2005 level.[11] On the evidence to date, there are serious obstacles to scaling up this mechanism to generate the required resources for developing countries. The need for effective regulation and monitoring of innovative financial instruments such as CDM projects may raise administrative costs and deter their use, particularly in developing countries. Major investments in training and education are also likely to be required. The success of the sulphur trading scheme in the United States, often held up as a model for emissions trading, appears to have depended on a number of complementary initiatives and investments (see box VI.1); the same will likely be true for carbon trading.

From a development perspective, cap and trade is problematic because, by allowing richer countries to use credits from projects abroad to help meet their emission reduction targets, it allows those countries to avoid necessary mitigation efforts at home. The cap-and-trade system has been designed to conform to the policy experience, institutional capacity and economic conditions of rich countries. By default, this provides significant advantages to them, as the baseline for allowance allocations is the current emissions of the high-emitting countries.

Carbon markets will doubtless continue to expand, but the pace and scale of that expansion is unlikely to be sufficient to overcome the financial constraint on a low-emissions development pathway for developing countries.

Carbon taxes

By increasing the cost of emissions to private parties in a more predictable manner than cap and trade, carbon taxes provide the opportunity to both raise public revenues and mitigate climate damage. Their possible advantage lies in the more predictable price impact and the ease of design and administration. On the other hand, they can provoke intense political resistance, even in mature economies. Hence, proposals such as, for example, the International Monetary Fund's 2008 call for a global tax on carbon have not yet demonstrated that they are politically feasible options.[12]

The United Nations Development Programme (UNDP) has estimated that a $20 tax per ton of CO_2 imposed on the member countries of the

Box VI.1: Sulphur trading and why it worked

Market mechanisms do not work in a vacuum: they are shaped by many factors. The United States system of sulphur emissions trading, the inspiration for many cap-and-trade proposals, is often credited with having triggered a dramatic reduction in the costs of pollution control. The Clean Air Act Amendments of 1990 had established the system, setting a cap on sulphur emissions at about half of the 1980 emissions and distributing allowances to businesses, roughly in proportion to past emissions. All large stationary sources of sulphur emissions, primarily coal-burning power plants, were included. The trading system was phased in from 1995 to 2000, with costs of controlling sulphur far below the levels that had been anticipated in advance.

However, this result cannot be attributed to trading alone: the low cost made itself apparent quite early, at a time when the volume of emissions trading was quite small. Several other events also played important parts in driving down the costs. Just before trading began, a sharp reduction in railroad freight rates made it affordable to bring low-sulphur coal from Wyoming to Midwestern power plants, replacing high-sulphur coal from the closer Appalachian coalfields. Some state regulations required even greater sulphur reduction than that stipulated by the national law, so it took no extra effort for power plants in those states to comply with the new national standard. At the same time, prices were declining for scrubbers, the pollution control devices that remove sulphur emissions. In this context, the emissions trading system may have made some contribution to lowering costs, but it operated on a field tilted in its favour. Without all the helpful coincidences, sulphur emissions trading would have looked much less successful.

If the United States sulphur emissions trading experience is the model for a carbon market mechanism, then the most important question may be, what other initiatives are needed to complement the market and again tilt the field in favour of success? It is not hard to identify the areas—energy efficiency, and low-carbon and no-carbon energy sources—where investment in research and development are needed. This is not just a matter of costs, but also of opportunities—to create new industries and jobs and to launch a promising new path of technological development.

Source: Ackerman, 2009.

Organization for Economic Cooperation and Development (OECD), at current emission levels, could produce $265 billion.[13] Many OECD countries already have carbon taxes, used primarily as general revenue sources, not to finance low-emissions development. The EU also applies differential taxes on energy to products, such as natural gas compared with diesel or petrol, when they are used as motor or heating fuel. These taxes appear to have contributed to energy efficiency, but they have hardly been sufficient to counter the threat of warming temperatures.

Other schemes have been proposed to specifically finance climate change activities. A small per-passenger levy on international flights, for example, would not only raise revenue, but because air fuel is often tax-exempt, it would reduce the implicit subsidy for air travel relative to other modes of transportation. Reducing subsidies to fossil fuels could help lower emissions and provide incentives for the transition towards a low-emissions economy. Subsidies to petroleum fuels—the difference between the end-user price and the price in a competitive market—have been estimated at $300 billion per year or 0.7 per cent of world GDP.[14] But particularly in developing countries, raising the price of essential goods such as energy could render them unaffordable by lower income groups, an outcome that would be widely viewed as socially unacceptable.

Any global carbon tax would require multilateral cooperation to harmonise tax systems, jointly set the level and incidence of the tax, and allocate the revenues. Without a robust international framework, differentiated taxes may serve discriminatory political or trade objectives instead of furthering climate change mitigation. Still, carbon taxation could create incentives towards mitigation in advanced countries and revenue for climate programmes, including in developing countries. With a carbon price of $50 per ton of CO_2, renewable energy such as onshore wind would be roughly competitive with dirty coal, and roughly $500 billion in resources could be mobilized annually[15]—though there would be competing priorities for the use of those funds. The potential revenue is not limitless, and would drop off as greenhouse gas emissions are reduced, but in the initial stages, carbon taxes may play an important role in climate finance.

Pricing ecosystem services

A related mechanism entails imposing fees and levies for activities and services whose benefits are not adequately captured by market prices. Owing to their specificity, ecosystem services cannot be traded as easily as liquid financial assets. As an alternative, several methodologies have been created to assess market value of these services and charge the potential beneficiaries, using shadow prices in a "pay as you use the service" approach. The idea of preserving ecosystems through the use of the services they provide is at the core of the strategies to reduce emissions from deforestation (see box VI.2).

However, setting and administering such a tax may be difficult for many developing countries. Estimates of damages caused by carbon emissions—the basis for the shadow price of carbon—also vary widely, because of the different assumptions made in order to value inter-temporal trade-offs or non-monetary damages, or to account for incomplete information and uncertainty.

Box VI.2: Reduction of emissions from deforestation and forest degradation (REDD)

In addition to providing many other products and services, forests can play a key role in tackling climate change. Deforestation, forest degradation and land-use changes produced 12.2 per cent of global greenhouse gas emissions in 2005, roughly the same as the transportation sector, and slightly more than all the European Union countries combined. Emissions from deforestation alone could increase atmospheric carbon stock by about 30 parts per million (ppm) by 2100. Thus, forest protection will need to form a central part of any global climate change deal.

Nicholas Stern and others have deemed curbing deforestation a cost-effective and relatively quick way of reducing emissions. Emissions from the forest sector could be cut in half by 2030 at an estimated cost of between $17 billion and $33 billion per year. This is far below the cost per ton of most emission reduction options in the energy sector.

At present, only a very small share of the existing investment in the forest sector is allocated to addressing climate change, and less than 25 per cent of that share is invested in developing countries and economies in transition. Fortunately, the importance of limiting deforestation and forest degradation has been recognized by climate change negotiators, and several new financing initiatives have been launched. By far the most significant is Norway's commitment to provide $600 million annually towards efforts to reduce carbon emissions from deforestation and forest degradation in developing countries. Other donors, including Australia, Finland, Spain, Japan, Switzerland, the United Kingdom and the United States, have contributed or have signalled their intent to contribute funds to climate change and forests programmes.

The World Bank has established the Forest Carbon Partnership Facility to help reduce emissions from deforestation and degradation and to help build capacity for REDD activities. Thirty-seven developing countries were participating as of mid-2011, and 16 financial contributors—governments as well as private entities—had pledged about $450 million for the programme. Just over half these funds are to support readiness reforms and investments identified through national REDD strategies; the rest will provide payments for verified emission reductions from REDD programmes in those countries.

Development of a REDD mechanism must be based on sound methodologies for estimating and monitoring changes in forest cover, associated carbon stocks, and greenhouse gas emissions, along with incremental changes due to sustainable management of forests, and reduction of emissions due to deforestation and forest degradation. The methodological challenge has proved to be more difficult for emissions due to forest degradation than for deforestation. Other policy issues to consider include the rights of forest stakeholders, in particular indigenous peoples, and the opportunity costs of other land uses and forest management systems. Negotiators should also ensure that REDD does not disadvantage countries that have already taken steps to eliminate or reduce deforestation and to manage their forests sustainably, or countries where forests are sustainably managed.

Sources: UN/DESA; United Nations Forum on Forests Secretariat; World Resources Institute, 2010; World Bank, 2011.

Sources of 'green' investment

Equity markets could provide another mechanism for mobilizing private investment in green technologies and low-emissions energy supply and infrastructure. Incentive structures would need to shift to favour such investments; this could be achieved if clear, aggressive reduction targets produced a carbon price that raised the profitability of low-emissions investments, or if fiscal incentives and public investments raised the rate of return on private-sector "green" investments. Foreign direct investment (FDI), portfolio investment, microfinance and public-private partnerships could be promoted to scale up private financing for climate change mitigation and adaptation.

Foreign direct investment (FDI) can be a relatively stable source of financing, with advantages in terms of transferring technology and standards which could allow for leapfrogging into cleaner sectors such as renewable energy. Some of the big emitting sectors, such as road transport, metals, mining, chemicals, timber, and cement are dominated by large international firms. Their investments and practices will likely have a big influence on the timing of alternative development pathways. Moreover, given the advances in clean technologies made by some developing countries in, for example, wind technology, South-South FDI may be an important component of a new development pathway. However, since FDI tends to lag rather than lead economic growth, it is unlikely to play a significant role in the early stages of a shift onto a new development pathway—particularly given the initial high degree of uncertainty and the absence of the domestic inputs and complementary investments that large international firms often expect.

Portfolio investments may be mobilized through venture capital funds as well as "green" funds and stocks and could appeal to investors willing to accept lower returns to support mitigation and socially responsible business practices. However, the funds made available through this channel to developing countries so far have been both limited and skewed in favour of one or two countries. Without a sufficient rise in the price of carbon and government intervention through regulatory measures and fiscal incentives, the private sector will not find these instruments sufficiently attractive—especially when there is price volatility, as seen recently with ethanol.

Nonetheless, some private equity investment firms are beginning to perceive clean infrastructure, primarily renewable energy, as offering viable financing opportunities. Activity is limited by several factors, including a lack of infrastructure to support production and distribution of renewable energy. In China, the largest market for this type of private financial flow, there are challenges to private investment because of national policies requiring links with domestic firms. Nevertheless, investment banks are increasingly seeing opportunities, most likely because of renewable energy quotas and feed-in tariffs that reward investment in this area, and investors are beginning to act on these prospects. Again, this trend underscores the need for rapid action in policy creation; private investors, particularly in this market, may take significant time to respond to incentives.

Microfinance could be another vehicle for mobilizing local private resources for investments in sustainable development. Over the past three decades, microfinance has grown dramatically around the world, even in some developed countries, with thousands of institutions serving more than 100 million people. In China alone, there were more than 2,600 microcredit companies at the end of 2010, up from 500 in 2008.[16] Microfinance has expanded beyond the original programmes of credit provisioning and now includes schemes of microsavings and microinsurance. Given the close links between poverty reduction and climate vulnerability, scaling up microfinance is a possible source of finance for climate adaptation. The Grameen Bank has already begun to extend loans for clean energy products, such as solar home systems, with spin-offs to microenterprises, while further opportunities exist in cleaner cooking products, biofuels and low-emissions agriculture. However, scaling up microfinance for long-term investment in productive activities and sustainable development will require support through a broader development strategy, including investments in infrastructure and human capital.[17]

Public-private partnerships and guarantees can help stimulate private financing in projects for increasing energy efficiency and renewable energy in developing countries. Partnerships have assumed growing importance in recent years as a vehicle for infrastructure projects and delivery of health services. They have also been used to bolster technological development, including in the field of clean energy. However, there are doubts about their cost-effectiveness and whether they represent the best way to deliver at scale.

Guarantees can take various forms. In southern India, for example, a consumer financing programme for photovoltaic systems helped consumers make the necessary upfront investments to use solar energy. The International Finance Corporation (IFC), the private sector arm of the World Bank Group, has been particularly innovative in this area. By establishing partnerships with banks in developing countries, IFC helps local financial institutions identify which of their clients could implement energy efficiency programmes. When a loan is given, training is provided on how to structure those programmes to further encourage investments, and IFC also issues a partial risk guarantee against default. With default rates actually lower than in other sectors, the guarantees and training have thus made efficient use of IFC resources, helping the private sector overcome its initial reluctance to invest in energy efficiency and renewable energy sectors in developing countries.[18]

Public sector financing

In many developing countries the markets for long-term financing, such as bond markets, are weakly developed. This typically limits what both governments and private investors can do to mobilize long-term capital for large-scale investments in economic and social development. The costs may be too large for governments to finance from yearly tax revenue, and the lack of a bond market limits the capacity for domestic public borrowing for these purposes. Investors, in turn, will anticipate private returns below social returns in the investments concerned.

Economy-wide externalities are particularly prominent in certain key sectors, such as infrastructure, which are characterized by lumpy investments, long gestation lags, higher risks and lower profits. Market signals can result in the avoidance of these sectors by investors, thereby slowing long-term growth and development. In order to overcome this obstacle, policymakers must ensure an adequate flow of credit at favourable costs to frontline technologies and sectors with potentially large social returns. To accelerate private investment in mitigation, policymakers and public authorities will need to apply incentives through regulatory frameworks, subsidies, guarantees, and financing of the incremental costs of switching technology, among other policy instruments.

Domestic resource mobilization

In the logic of a big push, increased public investment creates a matching amount of *new* saving, instead of drawing on *existing* saving. That higher level of saving, in turn, creates demand for new financial instruments, including the funding of public sector investments. However, this does not occur automatically, and financing gaps have to be filled. Governments must consider how their fiscal space can be expanded and refocused in order to meet their climate objectives without jeopardizing other goals. This is true of developed and developing countries alike, but the challenge is particularly significant in the latter.

In developing countries, on average, the tax revenue collected as a proportion of GDP is only two thirds of the proportion in richer countries, and a larger share is in the form of indirect taxes, as opposed to direct taxes on incomes, profit and capital gains. Tax administration is often weak and subject to evasion and abuse.[19] In identifying the resources needed for low-emissions growth, developing countries should undertake fiscal reforms that enable a shift away from a reliance on trade and other indirect taxes, with a view to increasing progressivity and expanding the fiscal space.

On the expenditure side, many governments are being advised, on climate change-related grounds, to reconsider energy subsidies for low-income households. While removing energy subsidies for low-income households would have a narrowly defined fiscal benefit, both the climate impact and the single-minded focus on this subsidy are questionable. Faced with higher energy prices, low-income households have been known to substitute unpriced energy sources, such as firewood, with a negative impact on the environment and on their own productivity and standard of living.

Low-emissions financing strategies will require an ensemble of subsidies, tariffs and taxes, of which energy subsidies for the poor should be only a part. On the revenue side, equity considerations will also have to play a major role in generating the needed financing for low-emissions energy investment, and progressive approaches to taxation and fees should be key elements in a climate financing strategy.

A number of developing countries have witnessed growth in markets for government bonds in recent years. Issuance of "green bonds" to fund climate change programmes could be an additional financing tool, along the lines of war bonds, in some emerging economies, and they could be an attractive, safer haven for personal savings.[20] Government guarantees and tax breaks could also be used to channel savings into investments that reduce carbon use, including infrastructural investment, as is done in some instances in the United States municipal bonds market.

The scale on which "green" debt instruments can be issued depends in part on the sophistication of domestic financial markets and the overall debt burden of the country. Expansion of a market for such funds is ultimately contingent on the national government's ability to raise tax revenues and to set the rate of return on domestic investment. Equity and development considerations are important on both fronts.

Public sector development banks provide an alternative funding channel for long-term investment in many developing countries. These institutions' record is uneven, although they have played a particularly important role in infrastructure development. At their best, they can encourage the development of complementary private financial institutions, and avoid excessive public sector risks and badly targeted interest-rate subsidies. In recent years, these institutions have been neglected in favour of private capital markets and public-private partnerships. However, in the absence of effective regulatory, policy and institutional frameworks, the private sector's record has not been satisfactory, particularly in financing essential utilities and services such as energy. In many cases, reforming and recapitalizing development banks will be important for a successful transition to low-emission development pathways.

International financing

The high-growth, low-emissions pathway described in this book will move developing countries toward self-sufficiency in financing investment, more rapidly than a conventional market-based scenario. In order to launch the developing world on that pathway, however, international support is indispensable for financing a "big push" of public investment in mitigation and adaptation.

International negotiations have agreed to provide such support on a number of occasions, but the reality has fallen far short of what is needed. At the UNFCCC negotiations in Cancún in 2010, the Parties agreed to establish a Green Climate Fund (GCF) that is expected to play a major role in international climate mitigation and adaptation finance, but they left its design, funding levels and governance to future negotiations.[21] To date, the primary climate financing sources under the UNFCCC have been two grant-based adaptation funds administered by the Global Environment Facility (GEF), which rely on voluntary contributions, and the Adaptation Fund under the Kyoto Protocol, which became operational in 2009 and is primarily funded by a 2 per cent levy on transactions under the CDM. The GEF funds, which had disbursed a combined $244 million to climate-related projects as of March 2011, have been particularly important because they can fund more risky projects and have demonstrated competence in working in countries that may not attract foreign investors either through the CDM or directly.[22]

A second channel encompasses funds and programmes arising from the loans and grants of bilateral agencies, the largest of which is Japan's Hatoyama Initiative, which has pledged $15 billion over five years, including $4 billion from private sources.[23] As noted above, bilateral entities have been the biggest funders of climate initiatives in recent years, with the top four entities providing $13 billion in 2009 alone.[24]

The third channel comprises existing multilateral development institutions, which include a variety of mechanisms with a climate-related component, and have set up several specific funds to provide loans, grants and concessional funding. These include the Climate Investment Funds of the World Bank, a $6 billion multilateral initiative announced at the 2008 G8 meeting.

This emerging climate architecture is as unnecessarily complex as it is woefully underfunded. The array of funds and funding mechanisms lack adequate coordination, leaving many gaps and overlaps, and there is great uncertainty about the total level of funds that developing countries can expect to receive. In any case, the World Bank/UNDP Climate Finance Options platform notes that "current international funding dedicated to climate action in developing countries covers only perhaps 5 per cent of their anticipated needs."[25]

Scaling up climate financing will require finding more predictable multilateral sources. These could come, in part, from the sale of emissions permits or increased carbon taxes in donor countries, but more innovative sources will likely be needed. One option would be a joint, wide-ranging initiative to mobilize countries of widely varying situations to achieve internationally agreed development goals, with new, predictable financing mechanisms. A hallmark of this approach is global solidarity, with sources of finance coordinated internationally but implemented at a national level. Unlike traditional development financing approaches, which still depend on the political goodwill of rich countries, this would entail joint design and decision-making by developing and developed countries.

A global investment regime for the climate challenge

While market-based approaches will need to be part of the solution, the key focus of a wider approach should be on making major investments to simultaneously address climate change, sustainability and economic development. Without significant financial transfers from wealthy countries, any expectation that poorer countries will move onto a low-emissions growth path is almost certain to be disappointed.

This investment-led approach seeks to change countries' development trajectory so they can meet their goals while reducing their carbon dependence. At the national level, and as part of a long-term industrial development strategy, public investment in mitigation and adaptation activities needs to be scaled up. Energy provision is a central component of this strategy, but it is interlinked with transportation, water security and economic diversification (chaps. II and III). Industrial policy—understood not only as targeting and coordinating specific sectoral support measures, but also socializing investment risks, removing barriers to adopting otherwise profitable technologies, and supporting technological learning and upgrading—has a key role to play both domestically (chap. IV) and internationally (chap. V). A successful investment push along these lines would in turn increase productivity and reduce the costs of using new technologies, opening up further investment opportunities.

In comparison with conventional market-based mechanisms, which would likely be accompanied by structural adjustments, a globally funded public investment programme would promote equity by enabling the developing world to sustain catch-up growth through the mobilization of domestic resources, while making significant cuts in emissions (chap. I). Such an investment programme would utilise market mechanisms insofar as government policy provided clear signals to private enterprises about the next wave of investment opportunities, without relying on a single price-based intervention.

Efforts to develop an investment programme that combines development and environmental goals on such a large scale have been few and far between. One recent example of the success of such efforts, even more telling since it has been achieved by a developing country, is the establishment by Brazil of a sugar cane-based ethanol energy and transport system (see chap. IV, box IV.3). A historical example concerns an underappreciated component of New Deal policies in the United States of the 1930s, the Tennessee Valley Authority (TVA; see chap. IV, box IV.1). With support from the Rural Electrification Administration and the Reconstruction Finance Corporation, the TVA transformed the Southern states by lowering transport costs, reducing the risk of flooding and creating a low-cost source of clean energy (hydroelectric power) that not only directly raised living standards but also helped the region crowd in substantial private investment and create new jobs. What is needed this time around is a global TVA: the investment programme that can meet the climate challenge must be a truly worldwide project.

Elements of a global programme

Estimates of mitigation and adaptation costs suggest that up to $1 trillion per year in new investments could be needed in developing countries to be mobilized from domestic and international resources. The breakdown between the public and private sector will no doubt vary considerably across time and among countries. However, the initial push will depend heavily on the public sector, reflecting the need to front-load much of the required investment.

In this context, it is clear that there needs to be a radical shift in the existing system of funding for mitigation and adaptation efforts, including an expanded public investment component. A global approach to a publicly funded investment programme should be based on three elements:

- A development accord that recognizes equity as an integral part of a global response to climate change.
- Additional and substantially scaled-up financing to allow for climate action with greater urgency.
- Independent and participatory governance structures along the lines of the Marshall Plan.

A development accord

Equity is an essential ingredient of an effective global climate change policy, as reflected in the principle of "common but differentiated responsibilities and respective capabilities" in the UNFCCC. Not only have today's high-income economies generated about 80 per cent of past fossil fuel-based emissions, but those same emissions have helped carry them to high levels of social and economic well-being. These countries carry the responsibility for the bulk of climate damage, but they also have the capacity to repair it. However, from a long-term perspective, limiting further damage also requires that developing countries shift their energy and land use and their consumption needs towards low-emissions options.

Compelling developing countries to cut their emissions at this stage, relying only on their own resources, would be inappropriate and unworkable; it would almost certainly freeze a pattern of income inequality that already exhibits intolerably wide gaps within and across countries. Catch-up growth and convergence remain fundamental policy priorities. Reconciling this with climate objectives is only possible if the investments that drive future growth assume a technological profile different from the one that drove the past growth of today's advanced economies.

Some developing countries are already taking significant steps towards energy efficiency and cleaner energy sources, and are building multilateral support to finance further emissions reductions at an accelerated pace.

However, much greater investments are needed for global progress at the right pace and scale to meet both climate and development goals.

This will require additional multilateral financing, on an adequate and predictable scale, comprising grants, concessional loans and compensatory payments. In UNFCCC negotiations, developing countries have insisted that Annex II (high-income) countries have a clear-cut responsibility to provide additional financial resources to meet the full costs incurred by developing-country parties in complying with their obligations. Translating such responsibilities into tangible resources has been a major stumbling block, but the need will only grow over time.

Additional and substantially scaled-up financing

The existing model of official development assistance (ODA) is not up to the task of funding the climate challenge. More substantial and more predictable forms of financing will have to be found, and new mechanisms of resource mobilization will have to be considered.

An obvious starting point would be to insist that advanced countries meet their longstanding commitment to a target 0.7 per cent of GNP for ODA.[26] Developing countries have rightly expressed concerns both about treating climate commitments simply as aid and about the potential for climate-related funds to "crowd out" assistance for development goals. However, given the links between climate vulnerability and other development challenges, there is extensive scope for synergies if developed countries remain faithful to their ODA commitments. There is no shortage of institutions available to channel such funding, but new funding mechanisms may still be needed, for example, in the area of disaster management. The bigger challenge is likely to be one of coordinating the required expansion of ODA, ensuring consistency across funding sources, and reducing duplication and waste. This may require the establishment of a central agency to coordinate international adaptation funding and to provide some degree of coherence across programmes.

Problems with aid governance will also need to be addressed. The lack of transparency in the donor-dependent approach to the design of specific-purpose funds needs to be corrected; international cooperation should help integrate mitigation and adaptation in the national policies of developing countries under the "country-led and country-owned" principle. There is also a need to rationalise and minimize the proliferation of funding mechanisms. There are numerous specific funds administered by bilateral agencies, which differ widely in purposes, size, time horizons and funding mechanisms. The "bilateralization" of multilateral aid should be minimized by coordination and integration of resources; for example, funding for reducing emissions from deforestation and forest degradation could expand by combining resources and approaches from different institutions.

Independent and participatory governance structures

At a time when international unity and coordination are needed, donor governments seem to have opted for a disjointed approach to climate change, at the expense of effectiveness, efficiency and equity. A global investment programme aimed at effectuating the shift to low-emissions, high-growth development pathways requires a governance structure that is able to pursue a focused and coherent agenda, prevents dominance by donor countries and provides for participatory decision-making on financial contributions and disbursements. Nicholas Stern has argued, on these grounds, that the climate challenge needs a new institutional architecture.[27]

One major question concerns the management and allocation of financial resources. It is often argued that the World Bank and other multilateral development banks might be better positioned to scale up financing than a fund under the authority of the UNFCCC. However, these institutions have major limitations in the context of global environmental finance. For instance, the Climate Investment Funds administered by the World Bank have been criticised not only for their governance structure, which replicates the existing asymmetries of the Executive Board of the World Bank, but also for undermining the UNFCCC and for not being truly additional to existing ODA commitments.[28] Indeed, on their own assessment, multilateral development banks still are not systematically factoring climate change into their investment choices, and need to do more to ensure that all of their investments and lending operations take climate change into account.

Developing countries have also pointed out the injustice of having to borrow additional funds to switch to cleaner energy sources to address climate change, even though they have relatively little responsibility for the problem. This raises long-standing concerns for many developing countries about the role of development finance, including the privileged position of creditors in international financial negotiations, and the use of adjustment lending, through attached conditionalities, to constrain their policy options across a broad range of economic and social issues. They are concerned that housing any new financing mechanisms in the international financial institutions would subject them to the same governance arrangements and conditionalities as have been imposed on previous loans from these institutions.

The creation of the Green Climate Fund only partly addresses these issues: on one hand, it is to be "accountable to and under the guidance of the Conference of the Parties," with a 24-member board with equal representation for developing and developed countries. However, the World Bank has been chosen as the interim trustee, with a review after three years—a contentious decision opposed by several developing countries.[29] The establishment of the Green Climate Fund could still be an important first step towards the development of a broader institutional structure on global climate change financing. However, it also runs the risk of locking new financing into a donor-dominated, project-based approach, which would run counter to the arguments presented in this chapter.

Living up to the challenge: lessons from the Marshall Plan

Whatever the institutional details finally agreed to, the right model for meeting shared global challenges is still the Marshall Plan, as noted by Al Gore in his Nobel Lecture in 2007. The scale and urgency of the climate and development challenges warrant that kind of integrated emergency response. Moreover, part of the success of the Marshall Plan was that it bypassed the fledgling Bretton Woods institutions, which did not appear to be up to the job of fashioning policies and supporting institutional reforms attuned to local conditions. However, the Marshall Plan is not a blueprint which can simply be rolled out to meet contemporary challenges. Rather, it encompasses a set of broad principles which can be tailored to new circumstances and sensitivities. Despite the demonstrated success of the broad Marshall Plan framework in Europe in the 1940s, "aid" has developed over the years into a narrower mixture of assistance for specific projects and ad hoc responses to unexpected shocks, with little apparent coherence. Donor conferences are driven more by what donors want to promote than by the desire to support specific multi-year national programmes. It is difficult to see how aid can ever be really effective without an articulation of macroeconomic objectives and detailed programmes for infrastructure investment, a coherent account of priorities, and a sense of the necessary complementarities among different investments and projects.

National development programmes along the lines of the Marshall Plan would make it easier to provide general, non-project assistance to governments or for financing the balance of payments. This was crucial for a number of European countries under the Marshall Plan. The need for assistance to deal with long-term imbalances is usually seen by international financial institutions as evidence of a weak commitment to reform. This was not the view of the Marshall Plan authors, who regarded such assistance as an investment in structural change and as providing governments with the breathing space required to ensure the success of difficult and often painful policies. Today, the structural changes implied by the shift to a low-emissions development pathway will likewise cause fiscal and current-account pressures even as long-run adjustments are realized. The same wisdom that informed the Marshall Plan should apply to climate and development.

Another major attraction of a Marshall Plan framework is that it can serve an important political function. A multi-year programme for achieving economic and environmental objectives, setting out their interrelationships, the means to achieve them and their dependence on outside assistance, effectively embodies a government's vision of the kind of societal structure at which it is aiming. Highly political in nature, a proposed programme would provide the basis for democratic discussion and negotiations. The task is not an easy one, but obtaining popular support for such a programme can be a major stimulus for change. This will not always result in what the international financial institutions regard as the "best" policies, but the advantage of democratic processes is that they generate pressures to correct mistakes.

The creation of a "new Marshall Plan" could thus be the means of providing a concrete operational basis for such ideas as "ownership" and "partnership", which otherwise risk degenerating into empty slogans. Moreover, a coherent national programme bolstered by popular support, indicating where outside assistance could be most effective, becomes a powerful vehicle for persuading potential donors to respond to national priorities instead of imposing other agendas.

Conclusion

In terms of the need to secure international cooperation, the climate financing challenge is substantial and daunting. It is clear that, while market-based and voluntary approaches will have an important role to play over time, they are inadequate for meeting the immediate financing requirements. The shift to a low-emissions, high-growth development pathway in the developing world is unlikely to be led by private sector investment and risk-taking. Thus, more binding modalities of international cooperation must be pursued. The same limitations that hamper international cooperation in financing development apply to the response to climate change. In the face of this predicament, it is important to realize that the international community can overcome the two sets of limitations simultaneously by recognizing that a global investment programme directed towards climate change objectives also represents a key intervention in favour of development.

Notes

1 Zoellick, R., 2009. 'How Will the World Finance Climate Change Action?' Presented at the Bali Breakfast, Washington, D.C., 26 April. The World Bank. http://siteresources.worldbank.org/EXTCC/Resources/BBkfastFinance250409.pdf

2 United Nations Framework Convention on Climate Change, 2008. *Investment and Financial Flows to Address Climate Change: An update*. FCCC/TP/2008/7. http://unfccc.int/resource/docs/2008/tp/07.pdf

3 The global analysis is Margulis, S. and Narain, U., 2010. *The Costs to Developing Countries of Adapting to Climate Change: New Methods and Estimates – The Global Report of the Economics of Adaptation to Climate Change Study*. Report No. 55726. The World Bank, Washington, D.C. http://go.worldbank.org/JYCN0DKBZ0.

The regional study has been published as Narain, U., Margulis, S. and Essam, T., 2011. 'Estimating costs of adaptation to climate change.' *Climate Policy*, 11(3). 1001–19. doi:10.1080/14693062.2011.582387.

4 More recent estimates put the range somewhat higher, that is, between 2 and 3 per cent of world gross output (2010 values). For the lower bound estimate, see United Nations Environment Programme, 2011. *Towards a Green Economy: Pathways to Sustainable Development and Poverty Eradication*, Nairobi. http://www.unep.org/greeneconomy/GreenEconomyReport/tabid/29846/Default.aspx and, for the upper bound estimate, see United Nations, 2011. *World Economic and Social Survey 2011: The Great Technological Transformation*, New York. http://www.un.org/en/development/desa/policy/wess/wess_current/2011wess.pdf. It should be noted that both these studies refer not only to investment costs for climate change mitigation and adaptation, but also for sustainable food agriculture and forestry, water management, protection against natural hazards, and protection of biodiversity. Yet, in both cases by far most of the estimated cost relate to the green energy transformation. For these, as much as for the figures mentioned in the text, it applies that all these investment cost estimates are subject to significant uncertainty given assumptions that have to be made about future development and diffusion of new technologies and because of the interactions between all areas of sustainable development that are as yet not fully understood.

5 United Nations Environment Programme, 2010. *Bilateral Finance Institutions and Climate Change: A Mapping of 2009 Climate Financial Flows to Developing Countries*. Report prepared by the Stockholm Environment Institute and the UNEP Climate Change Working Group for Bilateral Finance Institutions. http://www.unep.org/pdf/dtie/BilateralFinanceInstitutionsCC.pdf

6 Smith, J. B., Dickinson, T., Donahue, J. D. B., Burton, I., Haites, E., Klein, R. J. T. and Patwardhan, A., 2011. 'Development and climate change adaptation funding: Coordination and integration.' *Climate Policy*, 11(3). 987. doi:10.1080/14693062.2011.582385.

7 Linacre, N., Kossoy, A. and Ambrosi, P., 2011. *State and Trends of the Carbon Market 2011*. The World Bank, Washington, D.C. http://siteresources.worldbank.org/INTCARBONFINANCE/Resources/State_and_Trends_Updated_June_2011.pdf

8 Nell, E., Semmler, W. and Rezai, A., 2009. *Economic Growth and Climate Change: Cap-And-Trade or Emission Tax?* SCEPA Working Paper 2009-4. Schwartz Center for Economic Policy Analysis, The New School, New York. http://ideas.repec.org/p/epa/cepawp/2009-4.html

9 United Nations Framework Convention on Climate Change, 2010. *The Contribution of the Clean Development Mechanism Under the Kyoto Protocol to Technology Transfer*. Bonn. http://cdm.unfccc.int/Reference/Reports/TTreport/TTrep10.pdf

10 United Nations Framework Convention on Climate Change, 2010. *Clean Development Mechanism Executive Board – Annual Report 2010*. Bonn. http://unfccc.int/resource/docs/publications/10_cdm_anrep.pdf

11 Linacre et al., 2011, op. cit.

12 International Monetary Fund, 2008. *World Economic Outlook: Financial Stress, Downturns, and Recoveries*. Washington, D.C. http://www.imf.org/external/pubs/ft/weo/2008/02/index.htm

13 United Nations Development Programme, 2007. *Fighting Climate Change: Human Solidarity in a Divided World*. Human Development Report 2007/8. New York. http://hdr.undp.org/en/reports/global/hdr2007-8/

14 United Nations Environment Programme, 2008. *Green Jobs: Towards Decent Work in a Sustainable, Low-Carbon World*. Report prepared by the Worldwatch Institute with technical assistance from the Cornell University Global Labor Institute. Nairobi. http://www.unep.org/labour_environment/features/greenjobs-report.asp

15 Stern, N., 2009. *A Blueprint for a Safer Planet*. Random House, New York.

16 See, for example, Reid, L. R., 2011. *State of the Microcredit Summit Campaign Report 2011*. Microcredit Summit Campaign, RESULTS Educational Fund, Washington, D.C. http://www.microcreditsummit.org/SOCR_2011_EN_web.pdf
Information on China is from People's Bank of China, 2011. 'PBoC Publishes 2010 Statistics on Microcredit Companies.' *PlaNet Finance*, 14 April. http://www.planetfinancechina.org/news/pboc-publishes-2010-statistics-microcredit-companies.

17 For a more in-depth treatment of this issue, see Rippey, P., 2009. *Microfinance and Climate Change: Threats and Opportunities*. Focus Note 52. Consultative Group to Assist the Poor (CGAP), Washington, D.C. http://www.cgap.org/gm/document-1.9.34043/FN53.pdf

18 This section is drawn from UN/DESA consultations with IFC staff.

19 Di John, Jonathan, 2007. 'The political economy of taxation and tax reform in developing countries.' In *Institutional Change and Economic Development*, H. J. Chang (ed.). Anthem Press and United Nations University Press, New York.

20 See Elliott, L., Hines, C., Juniper, T., Leggett, J., Lucas, C., Murphy, R., Pettifor, A., Secrett, C. and Simms, A., 2008. *A Green New Deal*. The Green New Deal Group. New Economics Foundation, London. http://www.neweconomics.org/publications/green-new-deal.

21 For more on design issues for the fund, see Bird, N., Brown, J. and Schalatek, L., 2011. *Design Challenges for the Green Climate Fund*. Climate Finance Policy Brief No. 4. Heinrich Böll Foundation, Washington, D.C. http://www.odi.org.uk/resources/download/5256.pdf

22 Global Environment Facility, 2011. *Status Report on the Least Developed Countries Fund and the Special Climate Change Fund*. Document GEF/LDCF.SCCF.10/Inf.2, prepared by The World Bank as trustee. Washington, D.C. http://www.thegef.org/gef/sites/thegef.org/files/documents/Status%20Report%20on%20the%20Climate%20Change%20Funds%20-%20May%202011.Rev_.1.pdf

Note that these figures are for the two UNFCCC funds administered by the GEF, and exclude the much-larger climate programme support provided through the GEF Trust Fund.

23 Heinrich Böll Foundation, 2011. 'Climate Funds Update.' 8 June. http://www.climatefundsupdate.org/

24 United Nations Environment Programme, 2010, op. cit.

25 The World Bank and United Nations Development Programme, 2011. 'Climate Finance Tracking.' *Climate Finance Options*. http://www.climatefinanceoptions.org/cfo/node/189. [Accessed 28 July, 2011.]

26 For a review of the history of the 0.7 per cent commitment and its importance in meeting development goals, see United Nations Millennium Project, 2006. 'The 0.7% target: An in-depth look.' http://www.unmillenniumproject.org/press/07.htm

27 Stern, N., 2009, op. cit.

28 See, for example, Tan, C., 2008. *No Additionality, New Conditionality: A Critique of the World Bank's Proposed Climate Investment Funds*. Bangkok Climate Change Talks Briefing Paper No. 5. Third World Network, Penang, Malaysia. http://www.twnside.org.sg/title2/climate/briefings/BP.bangkok.5.doc

29 Heinrich Böll Foundation, 2010, op. cit.

References

Ackerman, F., 2007. *Debating Climate Economics: The Stern Review Vs. Its Critics*. Report to Friends of the Earth-UK. Global Development and Environment Institute, Tufts University, Medford, MA. http://www.ase.tufts.edu/gdae/Pubs/rp/SternDebateReport.pdf

Ackerman, F., 2009. *Can We Afford the Future: The Economics of a Warming World.* London: ZED Books.

Ackerman, F. and Stanton, E.A., 2009. 'Projections Regarding Climate Change and Development.' Background paper prepared for United Nations *World Economic and Social Survey 2009.* http://www.un.org/en/development/desa/policy/wess/wess_bg_papers.shtml

Ackerman, F., Stanton, E. A., Hope, C. and Alberth, S., 2009. 'Did the Stern Review underestimate U.S. and global climate damages?' *Energy Policy*, 37(7). 2717–21. doi:10.1016/j.enpol.2009.03.011.

Agarwal, A. and Narain, S., 1991. *Global Warming in an Unequal World: A Case of Environmental Colonialism*. New Delhi: Centre for Science and the Environment.

Almeida, C., 2007. 'Sugarcane ethanol: Brazil's biofuel success.' SciDev.Net Spotlight: *Climate Change & Energy: The Biofuels Revolution*, 6 December. http://www.scidev.net/en/climate-change-and-energy/the-biofuels-revolution/features/sugarcane-ethanol-brazils-biofuel-success.html.

Amyris Brasil, 2011. 'Amyris to Supply São Paulo City Buses with Renewable Diesel from Sugarcane | Business Wire.' 19 July. http://www.businesswire.com/news/home/20110719005753/en/Amyris-Supply-S%C3%A3o-Paulo-City-Buses-Renewable.

Andersen, S.O., Sarma, K.M., and Taddonio, K., 2007. *Technology Transfer for the Ozone Layer: Lessons for Climate Change*. Earthscan, London.

Aniello, C., Morgan, K., Busbey, A. and Newland, L., 1995. 'Mapping micro-urban heat islands using LANDSAT TM and a GIS.' *Computers & Geosciences*, 21(8). 965–67. doi:10.1016/0098-3004(95)00033-5.

Azevedo-Ramos, C., 2007. *Sustainable Development and Challenging Deforestation in the Brazilian Amazon: The Good, the Bad and the Ugly*. Brazilian Forest Service, Ministry of Environment, Brasilia. http://www.fao.org/docrep/011/i0440e/i0440e03.htm.

Baer, P., Athanasiou, T. and Kartha, S., 2007. *The Right to Development in a Climate Constrained World: The Greenhouse Development Rights Framework*. Publication series on Ecology, vol. I. Berlin: Heinrich-Böll-Stiftung, Christian Aid, EcoEquity and the Stockholm Environment Institute. November. Available at http://www.boell.de/downloads/gdr_klein_en.pdf

Banerjee, L., 2007. 'Effect of Flood on Agricultural Wages in Bangladesh: An Empirical Analysis.' *World Development*, 35(11). 1989–2009. doi:16/j.worlddev.2006.11.010.

Barton, J.H., 2007. *Intellectual Property and Access to Clean Energy Technologies in Developing Countries: An Analysis of Solar Photovoltaic, Biofuel and Wind Technologies*. Trade and Sustainable Energy Series, Issue Paper 2. International Centre for Trade and Sustainable Development, Geneva. http://ictsd.org/i/publications/3354/.

Barton, J.H., and Maskus, K.E., 2006. 'Economic perspectives on a multilateral agreement on open access to basic science and technology.' In *Economic Development and Multilateral Trade Cooperation*, S.J. Evenett and B.M. Hoekman (eds.). World Bank and Palgrave/Macmillan, Basingstoke, U.K.

Bateman, F., Ros, J. and Taylor, J. E., 2009. 'Did New Deal and World War II Public Capital Investments Facilitate a Big Push in the American South?' *Journal of Institutional and Theoretical Economics*, 165(2). 307–41. doi:10.1628/093245609789273213.

Betts, R. A., Collins, M., Hemming, D. L., Jones, C. D., Lowe, J. A. and Sanderson, M. G., 2011. 'When could global warming reach 4°C?' *Philosophical Transactions of the Royal Society A: Mathematical, Physical and Engineering Sciences*, 369(1934). 67–84. doi:10.1098/rsta.2010.0292.

Bird, N., Brown, J. and Schalatek, L., 2011. *Design Challenges for the Green Climate Fund*. Climate Finance Policy Brief No. 4. Heinrich Böll Foundation, Washington, D.C. http://www.odi.org.uk/resources/download/5256.pdf

Blyde, J. S. and Acea, C., 2003. 'How does intellectual property affect foreign direct investment in Latin America?' *Integration and Trade*, 7(19). 135–52.

Brazilian Sugarcane Industry Association (UNICA), 2010. 'Sugarcane diesel-powered buses hit the streets of São Paulo.' 26 July. http://english.unica.com.br/noticias/show.asp?nwsCode={FEE3F6EA-1105-4112-B857-94D1010167B2}.

Butt, T. A., McCarl, B. A., Angerer, J., Dyke, P. T. and Stuth, J. W., 2005. 'The economic and food security implications of climate change in Mali.' *Climatic Change*, 68(3). 355–78. doi:10.1007/s10584-005-6014-0.

California Public Utilities Commission, 2011. *Renewables Portfolio Standard Quarterly Report—1st Quarter 2011*. Sacramento, CA. http://www.cpuc.ca.gov/NR/rdonlyres/62B4B596-1CE1-47C9-AB53-2DEF1BF52770/0/Q12011RPSReporttotheLegislatureFINAL.pdf

Cazenave, A. and Llovel, W., 2010. 'Contemporary Sea Level Rise.' *Annual Review of Marine Science*, 2(1). 145–73. doi:10.1146/annurev-marine-120308-081105.

Center for International Earth Science Information Network, Columbia University, 2007. 'National Aggregates of Geospatial Data: Population, Landscape and Climate Estimates, v.2 (PLACE II).' http://sedac.ciesin.columbia.edu/place/.

Chakravarty, S., and others, 2008. 'Climate policy based on individual emissions'. Princeton, New Jersey: Princeton Environmental Center, Princeton University.

Cline, W. R., 2007. *Global Warming and Agriculture: Impact Estimates by Country*. Center for Global Development & Peterson Institute for International Economics, Washington, D.C.

Commission on Growth and Development, 2008. *The Growth Report: Strategies for Sustained Growth and Inclusive Development*. The World Bank, Washington, D.C.

http://www.growthcommission.org/index.php?option=com_content&task=view&id=96&Itemid=169.

Cosbey, A. (ed), 2008. *Trade and Climate Change: Issues in Perspective.* Final Report and Synthesis of Discussions, Trade and Climate Change Seminar, Copenhagen, June 18–20, 2008. International Institute for Sustainable Development. http://www.iisd.org/pdf/2008/cph_trade_climate.pdf

Dell, M., Jones, B. F. and Olken, B. A., 2008. *Climate Change and Economic Growth: Evidence from the Last Half Century.* NBER Working Paper No. 14132. National Bureau of Economic Research, Cambridge, MA. http://www.nber.org/papers/w14132 .

Di John, Jonathan, 2007. 'The political economy of taxation and tax reform in developing countries.' In *Institutional Change and Economic Development*, H. J. Chang (ed.). Anthem Press and United Nations University Press, New York.

Dodman, D., Ayers, J. and Huq, S., 2009. 'Building resilience.' In The Worldwatch Institute, *State of the World 2009: Into a Warming World.* W.W. Norton and Company, New York. http://www.worldwatch.org/files/pdf/SOW09_chap5.pdf

Elliott, L., Hines, C., Juniper, T., Leggett, J., Lucas, C., Murphy, R., Pettifor, A., Secrett, C. and Simms, A., 2008. *A Green New Deal.* The Green New Deal Group. New Economics Foundation, London. http://www.neweconomics.org/publications/green-new-deal.

Enkvist, P.-A., Nauclér, T. and Rosander, J., 2007. 'A cost curve for greenhouse gas reduction.' *The McKinsey Quarterly*, No. 1. 35–45.

Epstein, P. R. and Ferber, D., 2011. *Changing Planet, Changing Health: How the Climate Crisis Threatens Our Health and What We Can Do About It.* University of California Press, Berkeley. 3.

Fan, Gang, and others, 2008. 'Toward a low carbon economy: China and the world'. Beijing, China: Economics of Climate Change. Draft paper.

Fieldman, G., 2011. 'Neoliberalism, the production of vulnerability and the hobbled state: Systemic barriers to climate adaptation.' *Climate and Development*, 2(2). 159–74. doi:10.1080/17565529.2011.582278.

Food and Agriculture Organization of the United Nations, 2010. *Global Forest Resources Assessment 2010.* FAO Forestry Department, Rome. http://www.fao.org/forestry/fra/en/.

Food and Agriculture Organization of the United Nations, 2008. *The State of Food and Agriculture, 2008.* Rome. http://www.fao.org/docrep/011/i0100e/i0100e00.htm.

Food and Agriculture Organization of the United Nations, 2004. *Trade and sustainable forest management: impacts and interactions.* Analytic study of the global project GCP/INT/775/JPN. Impact assessment of forest products trade in the promotion of sustainable forest management. FAO Forestry Department, Rome.

Gallagher, K.S., 2006. 'Limits to leapfrogging in energy technologies? Evidence from the Chinese automobile industry.' *Energy Policy*, 34(4). 383–94. doi:16/j.enpol.2004.06.005.

Gao, Guangsheng, 2007. 'Carbon emission right allocation under climate change'. Advances in Climate Change Research, vol. 3 (Supplement), pp. 87–91.

Gelman, R. and Kubik, M. (ed.), 2010. *2009 Renewable Energy Data Book*. National Renewable Energy Laboratory, U.S. Department of Energy, Washington, D.C. http://www1.eere.energy.gov/maps_data/pdfs/eere_databook.pdf

Global Commons Institute, 2008. 'Contraction and convergence: a global solution to a global problem.' Available at http://www.gci.org.uk/contconv/cc.html

Global Environment Facility, 2011. *The A to Z of the GEF: A Guide to the Global Environment Facility for Civil Society Organizations*. Washington, D.C. http://www.thegef.org/gef/AZ_CSO.

Global Environment Facility, 2011. *Status Report on the Least Developed Countries Fund and the Special Climate Change Fund*. Document GEF/LDCF.SCCF.10/Inf.2, prepared by The World Bank as trustee. Washington, D.C. http://www.thegef.org/gef/sites/thegef.org/files/documents/Status%20Report%20on%20the%20Climate%20Change%20Funds%20-%20May%202011.Rev_.1.pdf

Goldemberg, J., 2008. 'The Brazilian biofuels industry.' *Biotechnology for Biofuels*, 1(6). http://www.biotechnologyforbiofuels.com/content/1/1/6.

Goldemberg, J., Coelho, S. T., Nastari, P. M. and Lucon, O., 2004. 'Ethanol learning curve—the Brazilian experience.' *Biomass and Bioenergy*, 26(3). 301–4. doi:16/S0961-9534(03)00125-9.

Grubb, M., 2004. 'Technology innovation and climate change policy: An overview of issues and options. *Keio Economic Studies* (Japan), 41(2). 103–132.

Hansen, J., Sato, M., Kharecha, P., Beerling, D., Berner, R., Masson-Delmotte, V., Pagani, M., Raymo, M., Royer, D. L. and Zachos, J. C., 2008. 'Target Atmospheric CO2: Where Should Humanity Aim?' *The Open Atmospheric Science Journal*, 2. 217–31. doi:10.2174/1874282300802010217.

Heinrich Böll Foundation, 2011. 'Climate Funds Update.' 8 June. http://www.climatefundsupdate.org/.

Helm, D., 2008. 'Climate-change policy: Why has so little been achieved?' *Oxford Review of Economic Policy*, 24(2). 211–38. doi:10.1093/oxrep/grn014.

Hirschman, A.O., 1958. *The Strategy of Economic Development*. Yale University Press, New Haven, CT.

Hirschman, A.O., 1971. *Bias for Hope: Essays on Development and Latin America*. Yale University Press, New Haven, CT.

Hoekman, B. M., Maskus, K. E. and Saggi, K., 2005. 'Transfer of technology to developing countries: Unilateral and multilateral policy options.' *World Development*, 33(10). 1587–1602. doi:16/j.worlddev.2005.05.005.

Hufbauer, G.C. and Kim, J., 2009. 'Climate Policy Options and the World Trade Organization.' *Economics: The Open-Access, Open-Assessment E-Journal*, 29, 18 June. doi:10.5018/economics-ejournal.ja.2009.

Huq, S., and Osman-Elasha, B., 2009. 'The status of the LDCF and NAPAs.' Presentation at the International Scientific Congress on Climate Change: Climate Change: Global Risks, Challenges and Decisions (Copenhagen, 10–12 March 2009), session 41, *Adaptation to climate change in least developed countries: challenges, experiences and ways forward*. http://www.ddrn.dk/filer/forum/File/IARU_2009_Session_41_SaleemulHuq.pdf

Intergovernmental Panel on Climate Change, 2011. *IPCC Special Report on Renewable Energy Sources and Climate Change Mitigation*. O. Edenhofer, R. Pichs, Madruga, Y. Sokona, K. (coordinating lead authors); Seyboth, P. Matschoss, S. Kadner, T. Zwickel, P. Eickemeier, G. Hansen, S. Schlömer, and C. von Stechow (eds.). Cambridge University Press, Cambridge, U.K., and New York.

Intergovernmental Panel on Climate Change, 2007a. *Climate Change 2007: Synthesis Report*. Geneva. http://www.ipcc.ch/publications_and_data/ar4/syr/en/contents.html.

Intergovernmental Panel on Climate Change, 2007b. *Climate Change 2007: Impacts, Adaptation and Vulnerability. Contribution of Working Group II to the Fourth Assessment Report of the Intergovernmental Panel on Climate Change*. M. L. Parry, O. F. Canziani, J. P. Palutikof, P. J. van der Linden, and C. E. Hanson (eds.). Cambridge University Press, Cambridge, UK.

International Energy Agency, 2010a. *CO_2 Emissions from Fuel Combustion 2010—Highlights*. Paris. http://www.iea.org/co2highlights/co2highlights.pdf

International Energy Agency, 2010b. *World Energy Outlook 2010*. Paris. http://www.worldenergyoutlook.org/2010.asp.

International Energy Agency, 2008a. *Energy Technology Perspectives 2008: Scenarios & Strategies to 2050*. Paris. http://www.iea.org/textbase/nppdf/free/2008/etp2008.pdf

International Energy Agency, 2008b. *World Energy Outlook 2008*. Paris. http://www.worldenergyoutlook.org/2008.asp.

International Institute for Applied System Analysis (IIASA), 2009. *GGI Scenario Database*. Version 2.0. http://www.iiasa.ac.at/Research/GGI/DB/.

International Monetary Fund, 2008. *World Economic Outlook: Financial Stress, Downturns, and Recoveries*. Washington, D.C. http://www.imf.org/external/pubs/ft/weo/2008/02/index.htm.

Kindleberger, Charles, 1986. 'International public goods without international government.' *American Economic Review*, 76(1), 1–13.

Klare, M., 2008. 'Persistent energy insecurity and the global economic crisis.' Paper presented at the panel discussion on Overcoming Economic Insecurity, Second Committee, United Nations General Assembly, 11 November.

Kozul-Wright, R. and Rayment, P., 2007. *The Resistible Rise of Market Fundamentalism: Rethinking Development Policy in an Unbalanced World*. Zed Books and Third World Network, Penang, Malaysia. Chap. 4.

Loughry, M., and McAdam, J., 2008. 'Kiribati: relocation and adaptation.' *Forced Migration Review*, 31. http://www.fmreview.org/FMRpdfs/FMR31/51-52.pdf. 51–52.

Margulis, S. and Narain, U., 2010. *The Costs to Developing Countries of Adapting to Climate Change: New Methods and Estimates—The Global Report of the Economics of Adaptation to Climate Change Study*. Report No. 55726. The World Bank, Washington, D.C. http://go.worldbank.org/JYCN0DKBZ0.

McKinsey & Company, 2009. *Pathways to a Low-Carbon Economy: Version 2 of the Global Greenhouse Gas Abatement Cost Curve*. http://www.mckinsey.com/clientservice/ccsi/pathways_low_carbon_economy.asp.

Miller, B.A. and Reidinger, R.B., 1998. *Comprehensive River Basin Development: The Tennessee Valley Authority*. World Bank Technical Paper No. 416. The World Bank, Washington, D.C.

Miyamoto, K., 2008. 'Human capital formation and foreign direct.' *OECD Journal: General Papers*, 2008(1). 1–40. doi:10.1787/gen_papers-v2008-art4-en.

Moore, S., 2011. 'Strategic imperative? Reading China's climate policy in terms of core interests.' *Global Change, Peace & Security*, 23(2). 147–57. doi:10.1080/14781158.2011.580956.

Moreira, J.R., 2006. 'Brazil's experience with bioenergy.' In *Bioenergy and Agriculture: Promises and Challenges*, P. Hazell and R.K. Pachauri (eds.). International Food Policy Research Institute, Washington, D.C.

Murphy, J. M., Sexton, D. M. H., Barnett, D. N., Jones, G. S., Webb, M. J., Collins, M. and Stainforth, D. A., 2004. 'Quantification of modelling uncertainties in a large ensemble of climate change simulations.' *Nature*, 430(7001). 768–72. doi:10.1038/nature02771.

Nakicenovic, Nebojsa, 2009. 'Supportive policies for developing countries: A paradigm shift.' Background paper prepared for United Nations *World Economic and Social Survey 2009*. http://www.un.org/en/development/desa/policy/wess/wess_bg_papers.shtml

Narain, U., Margulis, S. and Essam, T., 2011. 'Estimating costs of adaptation to climate change.' *Climate Policy*, 11(3). 1001–19. doi:10.1080/14693062.2011.582387.

Narain, S. and Riddle, M., 2007. 'Greenhouse justice: an entitlement framework for managing the global atmospheric commons.' In *Reclaiming Nature: Environmental Justice and Ecological Restoration*, J. K. Boyce and E. A. Stanton, eds. London: Anthem Press, pp. 401–414.

Nell, E., Semmler, W. and Rezai, A., 2009. *Economic Growth and Climate Change: Cap-And-Trade or Emission Tax?* SCEPA Working Paper 2009-4. Schwartz Center for Economic Policy Analysis, The New School, New York. http://ideas.repec.org/p/epa/cepawp/2009-4.html.

Organization for Economic Cooperation and Development, 2007. Patent Database. Paris. http://stats.oecd.org.

Osman-Elasha, B., Goutbi, N., Spanger-Siegfried, E., Dougherty, B., Hanafi, A., Zakieldeen, S., Sanjak, E., Atti, H.A. and Elhassan, H.M., 2008. 'Community development and coping with drought in rural Sudan.' In *Climate Change and Adaptation*, N. Leary, J. Adejuwon, V. Barros, I. Burton, J. Kulkarni and R. Lasco (eds.). Earthscan, London.

Patz, J. A., Campbell-Lendrum, D., Holloway, T. and Foley, J. A., 2005. 'Impact of regional climate change on human health.' *Nature*, 438(7066). 310–17. doi:10.1038/nature04188.

People's Bank of China, 2011. 'PBoC Publishes 2010 Statistics on Microcredit Companies.' *PlaNet Finance*, 14 April. http://www.planetfinancechina.org/news/pboc-publishes-2010-statistics-microcredit-companies.

Reid, L. R., 2011. *State of the Microcredit Summit Campaign Report 2011*. Microcredit Summit Campaign, RESULTS Educational Fund, Washington, D.C. http://www.microcreditsummit.org/SOCR_2011_EN_web.pdf

Riahi, K. and Nakicenovic, N. (eds.), 2007. *Greenhouse Gases—Integrated Assessment*, special issue of *Technological Forecasting and Social Change*, 74(7). 873–1108.

Riahi, K., Grübler, A. and Nakicenovic, N., 2007. 'Scenarios of long-term socio-economic and environmental development under climate stabilization.' *Technological Forecasting and Social Change*, 74(7). 887–935. doi:16/j.techfore.2006.05.026.

Richardson, K., Steffen, W., Schellnhuber, H. J., Alcamo, J., Barker, T., Kammen, D. M., Leemans, R., Liverman, D., Munasinghe, M., Osman-Elasha, B., Stern, N. and Wæver, O., 2009. *Synthesis Report: Climate Change: Global Risks, Challenges & Decisions*. Synthesis report for Climate Congress held March 3-5, 2009, in Copenhagen. Key Message 1. http://climatecongress.ku.dk/pdf/synthesisreport.

Rickerson, W., Bennhold, F. and Bradbury, J., 2008. *Feed-in Tariffs and Renewable Energy in the USA—a Policy Update*. Report sponsored by the Heinrich Böll Foundation, the North Carolina Solar Center, and the World Future Council. http://archives.eesi.org/files/Feed-in%20Tariffs%20and%20Renewable%20Energy%20in%20the%20USA%20-%20a%20Policy%20Update.pdf

Rippey, P., 2009. *Microfinance and Climate Change: Threats and Opportunities*. Focus Note 52. Consultative Group to Assist the Poor (CGAP), Washington, D.C. http://www.cgap.org/gm/document-1.9.34043/FN53.pdf

Riveras, I. and Winter, B., 2011. 'Brazil seeks to boost stagnant ethanol industry.' Reuters, 11 June. http://www.reuters.com/article/2011/06/06/us-brazil-ethanol-idUSTRE7550PX20110606.

Sathaye, J., Jolt, E. and De La Rue du Can, S., 2005. *Overview of IPR practices for publicly funded technologies*. Paper prepared for the United Nations Framework Convention on Climate Change Expert Group on Technology Transfer. Available at http://unfccc.int/ttclear/pdf/EGTT/11%20Bonn%202005/IPRandOtherIssuesAssociatedwithPubliclyFundedTech.pdf

Schlenker, W., Hanemann, W. M. and Fisher, A. C., 2005. 'Will U.S. Agriculture Really Benefit from Global Warming? Accounting for Irrigation in the Hedonic Approach.' *The American Economic Review*, 88(1). 113–25. doi:10.1257/0002828053828455.

Schlenker, W. and Lobell, D. B., 2010. 'Robust negative impacts of climate change on African agriculture.' *Environmental Research Letters*, 5(1). 014010. doi:10.1088/1748-9326/5/1/014010.

Smith, J. B., Dickinson, T., Donahue, J. D. B., Burton, I., Haites, E., Klein, R. J. T. and Patwardhan, A., 2011. 'Development and climate change adaptation funding: Coordination and integration.' *Climate Policy*, 11(3). 987. doi:10.1080/14693062.2011.582385.

Stanton, E. A., Ackerman, F. and Kartha, S., 2009. 'Inside the Integrated Assessment Models: Four Issues in Climate Economics.' *Climate and Development*, 1(2). 166–84. doi:10.3763/cdev.2009.0015.

Stern, N., 2009. *A Blueprint for a Safer Planet*. Random House, New York. 12–13.

Stern, N., 2006. *The Economics of Climate Change: The Stern Review*. Cambridge University Press, Cambridge, UK. http://www.hm-treasury.gov.uk/stern_review_report.htm.

Tan, C., 2008. *No Additionality, New Conditionality: A Critique of the World Bank's Proposed Climate Investment Funds*. Bangkok Climate Change Talks Briefing Paper No. 5. Third World Network, Penang, Malaysia. http://www.twnside.org.sg/title2/climate/briefings/BP.bangkok.5.doc.

Todo, Y. and Miyamoto, K., 2006. 'Knowledge Spillovers from Foreign Direct Investment and the Role of Local R&D Activities: Evidence from Indonesia.' *Economic Development and Cultural Change*, 55(1). 173–200. doi:10.1086/505729.

United Nations, 2007. *The International Development Agenda and the Climate Change Challenge*, Committee for Development Policy, 2007, Policy Note. New York: United Nations. http://www.un.org/en/development/desa/policy/cdp/cdp_publications/climate_07.pdf

United Nations Conference on Trade and Development, 2007. *The Least Developed Countries Report, 2007: Knowledge, Technological Learning and Innovation for Development*. UNCTAD/LDC/2007. Geneva. http://www.unctad.org/templates/WebFlyer.asp?intItemID=4314&lang=1.

United Nations Conference on Trade and Development, 2002. *Economic Development in Africa—From Adjustment to Poverty Reduction: What Is New?* UNCTAD/GDS/AFRICA/2. Geneva. http://www.unctad.org/templates/WebFlyer.asp?intItemID=2868&lang=1.

United Nations, 2011. *World Population Prospects: The 2010 Revision*. UN/DESA Population Division. http://esa.un.org/unpd/wpp/.

United Nations, 2008. *Climate change: Technology development and technology transfer*. Background document prepared for the High Level Conference on Climate Change: Technology Development and Technology Transfer, Beijing, 7 and 8 November 2008.

United Nations, 2011. *World Economic and Social Survey 2011: The Great Technological Transformation*, New York: Department of Economic and Social Affairs. http://www.un.org/en/development/desa/policy/wess/wess_current/2011wess.pdf

United Nations, 2010. *World Economic and Social Survey 2010: Retooling Global Development*, New York: Department of Economic and Social Affairs. http://www.un.org/en/development/desa/policy/wess/wess_archive/2010wess.pdf

United Nations, 2008. *World Economic and Social Survey 2008: Promoting Development, Saving the Planet*, New York: Department of Economic and Social Affairs. http://www.un.org/en/development/desa/policy/wess/wess_archive/2009wess.pdf

United Nations, 2006. *World Economic and Social Survey 2006: Diverging Growth and Development*. New York: Department of Economic and Social Affairs. http://www.un.org/en/development/desa/policy/wess/wess_archive/2006wess.pdf

United Nations and International Atomic Energy Agency, 2007. *Energy indicators for sustainable development: Country studies on Brazil, Cuba, Lithuania, Mexico, Russian Federation, Slovakia, and Thailand*. New York. http://www.un.org/esa/sustdev/publications/energy_indicators/full_report.pdf

United Nations Development Programme, 2007. *Fighting Climate Change: Human Solidarity in a Divided World*. Human Development Report 2007/8. New York. http://hdr.undp.org/en/reports/global/hdr2007-8/.

United Nations Environment Programme, 2011. *Towards a Green Economy: Pathways to Sustainable Development and Poverty Eradication*, Nairobi. http://www.unep.org/greeneconomy/GreenEconomyReport/tabid/29846/Default.aspx

United Nations Environment Programme, 2010a. *Bilateral Finance Institutions and Climate Change: A Mapping of 2009 Climate Financial Flows to Developing*

Countries. Report prepared by the Stockholm Environment Institute and the UNEP Climate Change Working Group for Bilateral Finance Institutions. http://www.unep.org/pdf/dtie/BilateralFinanceInstitutionsCC.pdf

United Nations Environment Programme, 2010b. *The Emissions Gap Report: Are the Copenhagen Accord Pledges Sufficient to Limit Global Warming to 2°C or 1.5°C?* http://www.unep.org/publications/ebooks/emissionsgapreport/.

United Nations Environment Programme, 2008. *Green Jobs: Towards Decent Work in a Sustainable, Low-Carbon World*. Report prepared by the Worldwatch Institute with technical assistance from the Cornell University Global Labor Institute. Nairobi. http://www.unep.org/labour_environment/features/greenjobs-report.asp.

United Nations Environment Programme and Bloomberg New Energy Finance, 2011. *Global Trends in Renewable Energy Investment 2011: Analysis of Trends and Issues in the Financing of Renewable Energy*. UNEP Division of Technology, Industry and Economics, Frankfurt School-UNEP Collaborating Centre for Climate & Sustainable Energy Finance, and Bloomberg New Energy Finance. http://fs-unep-centre.org/publications/global-trends-renewable-energy-investment-2011.

United Nations Environment Programme and GRID-Arendal, 2007. 'Coastal population and altered land cover in coastal zones (100 km of coastline).' *UNEP/GRIP-Arendal Maps and Graphics Library*. http://maps.grida.no/go/graphic/coastal-population-and-altered-land-cover-in-coastal-zones-100-km-of-coastline.

United Nations Framework Convention on Climate Change, n.d. 'National Adaptation Programmes of Action (NAPAs).' http://unfccc.int/national_reports/napa/items/2719.php. [Accessed 29 July, 2011.]

United Nations Framework Convention on Climate Change, 2010a. *Clean Development Mechanism Executive Board—Annual Report 2010*. Bonn. http://unfccc.int/resource/docs/publications/10_cdm_anrep.pdf

United Nations Framework Convention on Climate Change, 2010b. *The Contribution of the Clean Development Mechanism Under the Kyoto Protocol to Technology Transfer*. Bonn. http://cdm.unfccc.int/Reference/Reports/TTreport/TTrep10.pdf

United Nations Framework Convention on Climate Change, 2008. *Investment and Financial Flows to Address Climate Change: An update*. FCCC/TP/2008/7. http://unfccc.int/resource/docs/2008/tp/07.pdf

United Nations Framework Convention on Climate Change, 2007. *Climate Change: Impacts, Vulnerability and Adaptation in Developing Countries*. http://unfccc.int/resource/docs/publications/impacts.pdf

United Nations Human Settlements Programme (UNHabitat), 2008. *State of the World's Cities 2008/2009: Harmonious Cities*. Earthscan, London.

United Nations Human Settlements Programme (UNHabitat), 2007. *Global Report on Human Settlements 2007: Enhancing Urban Safety and Security*. Earthscan, London.

United Nations Millennium Project, 2006. 'The 0.7% target: An in-depth look.' http://www.unmillenniumproject.org/press/07.htm.

U.S. Bureau of the Census, 1975. *Historical Statistics of the United States, Colonial Times to 1970, Bicentennial Edition, Part 2*. U.S. Department of Commerce, Washington, D.C.

Vos, Rob and Kozul-Wright, Richard, eds., 2010. *Economic Insecurity and Development*, New York: United Nations (Sales E.11.II.C.3).

Winkler, H. (ed.), 2006. *Energy Policies for Sustainable Development in South Africa: Options for the Future*. Energy Research Centre, University of Cape Town, Rondebosch, South Africa. http://www.erc.uct.ac.za/Research/publications/06Winkler-Energy%20policies%20for%20SD.pdf

Winkler, H. and Marquand, A., 2009. 'Changing development paths: From an energy-intensive to low-carbon economy in South Africa.' *Climate and Development*, 1(1). 47–65. doi:10.3763/cdev.2009.0003.

World Bank, 2011. 'Introduction—Forest Carbon Partnership.' *The Forest Carbon Partnership Facility*. http://www.forestcarbonpartnership.org/fcp/node/12.

World Bank, 2008. 'Retracting Glacier Impacts Economic Outlook in the Tropical Andes.' Web article based on Vergara, W. et al., 2007, *Latin America: Results from the application of the Earth Simulator*. The World Bank. Washington, D.C. http://go.worldbank.org/PVZHO48WT0.

World Bank, Carbon Finance Unit, n.d. 'Colombia: Rio Amoya Run-of-River Hydro Project.' http://wbcarbonfinance.org/Router.cfm?Page=Projport&ProjID=54401. [Accessed 29 July, 2011.]

World Bank and United Nations Development Programme, 2011. 'Climate Finance Tracking.' *Climate Finance Options*. http://www.climatefinanceoptions.org/cfo/node/189. [Accessed 28 July, 2011.]

World Health Organization, 2010. *Climate and Health*. Fact sheet N°266. Geneva. http://www.who.int/mediacentre/factsheets/fs266/en/.

World Resources Institute, 2010. 'Climate Analysis Indicators Tool.' *CAIT 8.0*. http://cait.wri.org/

World Wind Energy Association, 2011. *World Wind Energy Report 2010*. Bonn. http://www.wwindea.org/home/images/stories/pdfs/worldwindenergyreport2010_s.pdf

Zoellick, R., 2009. 'How Will the World Finance Climate Change Action?' Presented at the Bali Breakfast, Washington, D.C., 26 April. The World Bank. http://siteresources.worldbank.org/EXTCC/Resources/BBkfastFinance250409.pdf

Index

F

H

I

M

N

O

V

W

PORTLAND PUBLIC LIBRARY SYSTEM
5 MONUMENT SQUARE
PORTLAND, ME 04101
WITHDRAWN